PROCESS ASSESSMENT
AND ISO/IEC 15504
A REFERENCE BOOK

THE KLUWER INTERNATIONAL SERIES IN ENGINEERING AND COMPUTER SCIENCE

PROCESS ASSESSMENT AND ISO/IEC 15504
A REFERENCE BOOK

by

Han van Loon
Consultant – SYNSPACE AG
Associate Professor
International University in Geneva, Switzerland
Visiting Professor
Nottingham Trent University, UK

Contributing Authors:
Ann Cass - *SYNSPACE AG*
Christian Steinmann – *HM&S*
Terry Rout, Angela Tuffley, Bruce Hodgen - *SQI*

 Springer

Cover illustration: © Han van Loon

Library of Congress Cataloging-in-Publication Data

A C.I.P. Catalogue record for this book is available
from the Library of Congress.

PROCESS ASSESSMENT AND ISO/IEC 15504:
A Reference Book by Han van Loon
Associate Professor
International University in Geneva, Switzerland
Visiting Professor
Nottingham Trent University, UK
Consultant – SYNSPACE AG
In association with SYNSPACE AG
Contributing Authors:
Ann Cass - SYNSPACE AG
Christian Steinmann – HM&S
Terry Rout, Angela Tuffley, Bruce Hodgen – SQI

ISBN 0-387-23172-2; e-ISBN 0-387-23173-0
The Kluwer International Series in Engineering and Computer Science
Volume 775
Printed on acid-free paper.

Printed in the United States of America.

9 8 7 6 5 4 3 2 1   SPIN 11054856, 11323983

springeronline.com

Dedication and Acknowledgements

This book is dedicated to the many professionals committed to improving their organizations through process assessment and quality improvement.

There are many people who have contributed to the development of the standards described herein, and I would like to acknowledge their collective contribution.

I would like to acknowledge that this book is based in part upon the work performed by many people in SYNSPACE AG beside myself, including Ann Cass, Christian Völcker and Hans Stienen. They have been instrumental in the extensive work on SPiCE for SPACE, RISK for SPACE and SPiCE for ISO 9000, and in providing input to the development of the ISO/IEC 15504.

In particular, I would like to thank Ann Cass who co-authored materials in the book in the sections on SPiCE for SPACE.

I would like to thank Terry Rout, Bruce Hodgen and Angela Tuffley from the Software Quality Institute for their work in Chapter 9 (with input from Christian Steinmann). Also many thanks for their many years of cooperation using ISO/IEC 15504, first while I worked at CelsiusTech and SAAB in Australia and later in Sweden, and the ongoing dialogue while at SYNSPACE.

I would also like to thank Lothar Winzer and Juan Maria Carranza from ESTEC (ESA) for their sponsorship of the creation of the SPiCE for SPACE document set, SPICE for ISO 9000, and permission to reuse parts of the work SYNSPACE has carried out for ESA.

I would also like to thank the following persons who contributed material; Daniel Keller on NOVE-IT and PAL, Christ Vriens for input on eXtreme Programming, Fritz Stallinger, John Torgersson and John Henderson-Sellers for papers on OOSPICE, and Gerhard Wagner for input on Automotive SPICE developments, Dave Kitson of the Software Engineering Institute for permission to use material about SW CMM® and CMMI® and Linda Ibrahim for material on the FAA iCMM.

I would like to thank the following people who provided reviews and feedback on various parts of the books: Ian Howgrave Graham, Ann Cass, Christian Völcker, Hans Stienen, Jürg Müller, Christian Steinmann, Bruce Hodgen, Terry Rout, Angela Tuffley, Ruth Seeger, Joachim von Linde, Gerhard Wagner, Perry Deweese, Daizhong Su, K. Beck and Fritz Stallinger.

Finally, I would like to also dedicate this book to my parents who taught me that striving for excellence brings its own rewards.

Han van Loon   Wednesday, May 12, 2004

TABLE OF CONTENTS

LIST OF FIGURES

LIST OF TABLES

Preface

Business is increasingly relying upon software, not only as a means to efficiently operate, but also as a means to effectively compete. More industries around the globe are using and creating software to differentiate their product and service offerings.

Software is increasing in functionality and complexity, and while this offers the opportunity to do more using software, it also creates more problems. The increasing problems associated with software reliability and complexity has been termed the software crisis, and it is costing business billions of dollars a year.

One important method to manage the impact of software, both for development and for use is via well-defined processes. Process assessment provides a means to assess how well an organisation performs these processes. ISO/IEC 15504 is emerging as the international standard for process assessment.

This is the first book dedicated to the standard. The book provides guidance for readers wishing to understand the power and benefits of a process approach and process assessment. It guides the reader through the various parts of the standard in an understandable and practical manner, grounded upon the people-process-product model.

Chapters 1 to 3 introduce the standard and basic concepts of processes and process assessment. Chapter 4 describes the measurement framework. Chapters 5 to 7 describe various process reference models, process assessment models and process lifecycle models including ISO standards and SEI CMMI® and the US FAA iCMM.

Chapter 7 also includes comparison to agile methodologies, especially Extreme Programming. Chapter 8 describes requirements for assessors and assessment teams. Chapter 9 provides a comparison to SEI CMMI®. Annexes provides the history of evolution of the standard and efforts towards harmonization of ISO standards, the proposed assessment model due in 2005 and a glossary of terms.

Process Assessment and ISO/IEC 15504: A Reference Book is structured to meet the needs of a professional audience composed of researchers, quality professionals and assessors in industry. This book is also suitable for graduate-level students in computer science, software engineering and MBA students. There is a companion book titled *Process Assessment and Improvement: A Practical Guide for Managers, Quality Professionals and Assessors.*

Introduction to the Process Assessment Reference Book

The Reference Book introduces the reader to the concepts of process assessment in general and the ISO/IEC 15504 process assessment standard in particular.

It has a companion guidebook: **Process Assessment and Improvement. A practical guide for managers, quality professionals and assessors.**

In the first chapter, I look very briefly at the creation of an international quality standard specifically addressing process assessment. I describe the foundations for organizational success in terms of a People-Process-Product model, the concepts of processes and process assessment in general, and how these are achieved in ISO/IEC 15504. It is written in a very general way and assessors, process and quality professional may wish to go to the more detailed sections suggested in the reading guide instead.

In chapter 2, an overview of the ISO/IEC 15504 standard and the structure of the document set are described, together with the reader's guidance to the standard. This provides the reader with a quick, high-level overview of the standard and its parts.

In chapter 3: Meeting the requirements of ISO/IEC 15504, I describe the overall requirements that the standard places upon assessment models and methods, and its role to harmonize across various standards, models and methods. This chapter covers the need for several levels of detail in process models by specifying the requirements for a conformant Process Reference Model that provides process purpose and outcomes, a conformant Process Assessment Model that can be used to assess the processes, plus how to map between the two models. The chapter also specifies the requirements for a conformant assessment in detail. Finally, the chapter discusses certification of assessors and assessment results.

In chapter 4, I explain the major dimension of process assessment: the capability dimension. The measurement framework describes the Capability Levels of ISO/IEC 15504, compares the new standard to the older Capability Levels in ISO/IEC TR 15504-2:1998, and describes the author's experience of the state of practice for each Capability Level. The way to rate process attributes using the NPLF rating scale is described. This is followed by the way process attributes are collated into process Capability Levels. Finally, I compare ISO/IEC 15504 to similar aspects of ISO 9000:2000.

In chapter 5, I explain the second dimension – the process dimension, as described in a Process Reference Model. The chapter describes several international standard models, particularly ISO/IEC 12207 and ISO/IEC 15288. The chapter then describes some industry and domain specific Process reference Models, including the Information Technology Infrastructure Library (ITIL ®) model, OOSPICE ® and SPICE for ISO 9000.

In chapter 6, I describe Process Assessment Models. The chapter starts by looking at the ISO/IEC 15504-2 requirements for Process Assessment Models, then the nature and use of process assessment indicators. To illustrate what a conformant Process Assessment Model contains, I describe the ISO/IEC TR 15504-5 exemplar, its structure and how compatibility, mapping and translation to the normative part of the standard are performed. I then summarize other conformant Process Assessment Models, including SPiCE for SPACE and its enhancements, OOSPICE for component-based development, and a brief introduction to the plans for Automotive SPICE. The SEI CMMI® and FAA-iCMM are also described. Finally, I describe the use of assessment indicators in rating processes.

In chapter 7, I describe some standard process lifecycle models. These models provide the next level of detail and guidance that an organization requires in order to implement a detailed process infrastructure. Three well-known models are described, the German V-Model, the Rational Unified Process and eXtreme Programming. The three models chosen illustrate the wide range of detail and approaches available to implement processes, and I give some guidance of how they relate to process assessment standards such as ISO/IEC 15504-2 and CMM®.

In chapter 8, I describe the requirements on assessors, specifically competence, both what is needed to gain the standing of competent assessor, as well as training, registration and certification. Finally, the chapter looks at how to assemble an assessment team that meets the assessor competence requirements of the standard.

In chapter 9, I summarize the SW CMM® and CMMI® relationship to ISO/IEC TR 15504:1998. This is mostly focused on CMMI® and compares its scope to ISO/IEC TR 15504:1998. The Continuous Representation Model Process Dimension and the Capability Dimension are described, including parts of their scope that are outside that of ISO/IEC TR 15504-2. Some issues concerning the Continuous Representation Model mapping and translation of assessment results are covered, and a representation of the mapping of the Staged Representation Model maturity model to the ISO/IEC 15504-2. This chapter also briefly describes CMMI® assessments. The popularity of CMMI® versus ISO/IEC 15504-2 and advantages and disadvantages of each are covered, including survey results for 2003, covering use and the recorded maturity levels of the surveyed organizations.

In Annex 1, there is a short history of process assessment development and the ISO/IEC 15504-2 harmonization aspects.

Annex 2 covers the proposed ISO/IEC CD 15504-5 Process Assessment Model.

Annex 3 has a Glossary of Acronyms and Terms.

Reader Guidance

The reader should keep in mind that parts of the standards described are still evolving and I have written the book against the best available information at the time or writing. Naturally I intend to keep up to date and active in the ISO/IEC 15504 community in order to be able to update future editions as necessary. I welcome your feedback on the book and also suggestions for topics to cover in the future.

The following table suggests the most useful chapters for readers of the reference book and the practical guide.

Reader Guidance to Books

Reader	Reader's interests and expected benefits	Reference Book	Practical Guide
Manager	Benefit from the use of process assessment and process improvement	1	1, 2, 5, 4, 9
Quality Professional	General quality system and process management. Comparing assessment models, and the uses of process assessment.	1, 2, 3, 4, 5, 7, 9, Annex 1	1, 2, 3, 4, 5, 8, 9
Assessors	Conducting a conformant assessment, developing the skills and competencies needed to perform an assessment. In-depth capability rating guidance.	3, 4, 6, 8, Annex 2, Annex 3	2, 6, 7, 8 Annex 1, Annex 2
Assessment Sponsor	How an assessment is conducted, what tools and other support are required, how to initiate an assessment. Various types of uses of assessments, Interpreting the results.	1, 3, 4, 8, Annex 3	1, 2, 3, 4, 5, 6, 8
Process Owner	Design and implementation of processes. Improving processes.	1, 5, 7	1, 5
Process Expert	Provide expertise on process design and applicability of the standard to design and improvement of processes.	2, 4, 5, 6, 7	5, 7, 8
Process Practitioner	Implement a process; participate in design and assessment of processes.		1, 2, 4, 5
Process Improvement Sponsor	Initiating an improvement programme, defining assessment inputs for an assessment for improvement purposes, using assessment results for improvement.	1, 3	1, 2, 5, 8, 9
Process Capability Determination Sponsor	Initiating a programme for the determination of supplier capability, defining a target capability profile, verifying and using assessment results in a capability determination exercise.	1, 3	1, 2, 4, 8, 9, 3 (optional)
Developers of Process Assessment Models	Developing Process Assessment Models for performing assessments against a compliant Process Reference Model and measurement framework of ISO/IEC 15504-2.	1, 2, 3, 4, 5, 6, 7 Annex 2, Annex 5	
Developers of Methods	Developing a method that will support the performance of conformant assessments.	3, 5, 6	
Tool Developers	Developing tools that support assessors by collecting, recording and classifying evidence in assessments.	3	Annex 1, Annex 2
Improvement Team Facilitator	Help teams in a Team Based Business Design Improvement process.		2, 5
Customer	Benefits and use of process assessment from a customer perspective, particularly for setting target profiles, capability determination and improvement.	1	1, 2, 3, 5, 8

1   YET ANOTHER QUALITY STANDARD?

Over the past three decades, there has been a proliferation of Quality Standards covering products and processes. These products and services are increasingly using software as a fundamental way to create competitive differentiation. The creation and use of software often requires more attention to the processes used to produce the products and services.

There are many organizations that develop these standards, including the International Organization for Standardisation (ISO for short), The International Electrotechnical Commission (IEC), the Institute for Electrical and Electronic Engineers (IEEE), and the United States government through bodies such as the National Institute for Standards and Technology (NIST). In addition, many other organizations are involved in contributing to standards development including the United Kingdom government through the UK Ministry of Defence (MoD) and the Software Engineering Institute (SEI).

There are industry standards both de facto and formalized by industry organizations, formal association driven standards, government standards, national standards and International standards.

There are standards that specify management approaches; standards that specify technical work approaches such as product development and testing; standards that specify safety and security requirements for products, standards that specify product compliance requirements; and standards that specify environmental management aspects.

In this book, I focus on standards related to processes, of which there are more than enough to keep anyone busy staying up to date with their evolution.

There are several standards addressing quality aspects of processes at international and national level.

Perhaps the most well known set of standards in quality are the ISO 9000 series.

ISO 9000 provides a general set of International Standards covering an organization's achievement of quality, especially quality from the customer viewpoint. The standard has evolved to focus more on development, implementation and improvement, in particular how it applies to the organization's processes for developing customer products and services.

When properly applied, standard such as ISO 9001 can help an organization to more efficiently and effectively produce quality products and services. When an organization applies ISO 9001, there is a requirement to assess how they perform. Traditionally ISO 9001 was used to audit an organization's processes for conformance[1].

So why another quality standard?

The international standards community decided that there was a need for a process assessment standard that provides for assessing *how well* processes are performed.

Another principal justification for a standard was that assessment approaches were multiplying, and they were not consistent and compatible with each other. Process assessment was being widely applied in commercially sensitive areas (supplier selection, supplier management) and incompatible approaches could lead to inconsistent results.

The international standards community therefore proposed a process assessment standard, created with the aim to consistently assess processes.

The standard consists of 3 main elements:

➢ a measurement framework,

➢ a set of requirements for defining processes, and

➢ a set of requirements for how to perform consistent assessments.

In addition, the standard should provide a means of harmonization between various international standards for processes and for professional organization process assessment methods. This would allow an organization to compare results from different conformant assessment methods; it allows them to *"compare apples against apples"*.

This standard is known as ISO/IEC 15504.

ISO/IEC 15504 differs from many preceding standards in two fundamental ways. First, it does not prescribe a way of working (i.e. the processes to be followed). Secondly, it does not focus on process compliance, but rather on how well processes are performed and managed.

It meets the needs of ISO 9001 sub clause 8.2.3 for monitoring and measurement of processes. In fact, ISO/IEC 15504 assessments have no

---

[1] In ISO 9004:2000 there has been a move towards assessing more than conformance, expanding it to assessing *how well* processes are performed.

'pass' or 'fail' criteria (compared to quality audits that traditionally assess for conformance or non conformance). Instead, ISO/IEC 15504 focuses on how well a process is performed, managed, defined, measured and improved.

1.1   Foundations for success

The preceding section emphasizes the existence of a large number of quality standards and this book focuses on process related aspects of these standards.

Let us just summarize: *why do we have quality standards?*

The basic reason for quality standards is to help organizations to achieve success by meeting their customer's needs, and to assure others that this achievement is as consistent as possible, so that they will benefit as customers (or partners or even suppliers).

So how do organizations achieve success?

Organizations rely on their people, processes and products to achieve success.

However people, processes and products interact in organizations in a variety of ways and no two organizations interact the same way. This is primarily because each organization has different people and groups of people. The people in the organization decide upon the products and processes.

In discussing process assessment, it is important to understand what a process is.

In ISO/IEC 15504, it uses the following definition:

Process – "*a set of interrelated activities, which transform inputs into outputs.*"

The ISO definition is mechanistic and a bit abstract– let's look at another definition.

The Cambridge Dictionary of American English definition is:

"*A process is a series of actions that you take in order to achieve a result.*"

This definition encompasses people, actions (processes) and results (outcomes). Furthermore, the results may be product oriented, people oriented or process oriented.

I model this interaction in a simple model I call the People-Process-Product model.

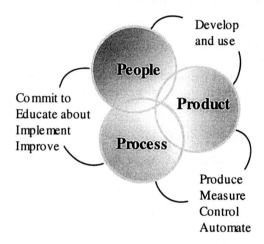

People-Process-Product model.

- People develop products (goods and services) for customers.
- People invent, commit to, educate themselves and others, and implement processes in order to produce these products.
- People use products (also often referred to as technology or tools) in their organization/business.
- Processes are used to measure and control the production of products.
- People improve processes as one way to improve products.
- Products can be used to simplify and automate processes.

Each organization can decide the balance between the way people, processes and products interact, the extent that they interact and the extent they rely on each. For example, I can decide upon various ways to achieve success:

- Do I allow people to choose the way they do something (e.g. craftsmanship)?
- Do I automate the way something is done by means of a product (e.g. an automated tool or software)?
- Do I define a standard process?

All three ways may be possible. Naturally, an organization striving for success wants to use the 'best way' (the most efficient and effective) that is available to them. This will normally require the organization to select a particular combination of people, processes and products that reflects the experience and knowledge of their people, the types of products they create and the processes they need to follow.

However, people vary in the way that they perform activities. This variation is a natural phenomenon. Different people can vary in doing the same or similar tasks, and one person can vary depending upon how they feel or think at different times. Sometimes this variation can be positive, for example finding ways to improve a product or service, sometimes this variation can be negative, for example failing to completely test a product due to time constraints. Poor processes can lead to poor or inconsistent quality products and failure. So appropriate processes are one of the 3 important organizational prerequisites to achieve success.

When people define and implement processes, they have an important means to manage achievement of their organizational success, and to manage variation in performance of the activities that achieve successful results.

When these people, processes and products interact together harmoniously and in synergy, they help the organization to achieve success.

1.2   A process view of the world

When people define and implement processes, they need to consider several aspects:

➢   What outcomes do I want to achieve?

➢   What outputs do I produce that achieve these outcomes?

➢   What inputs do I need?

➢   What are the activities I need to perform to create the outputs?

➢   Who needs to be involved?

➢   Do I need any other resources to support the activities?

➢   How do I control the process to make sure any variation does not cause poor quality?

➢   How do I check that I have achieved the right outcome?

The following diagram represents a simple process that covers all the above questions.

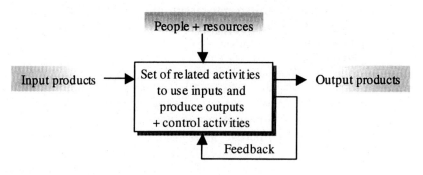

Simple process diagram

The process diagram not only has a set of activities focused on producing outputs that meet the desired outcomes, but also has a means to control variation (through use of feedback) that helps to manage the achievement of the desired outcomes.

It is possible to create one process that takes all the required inputs, uses all the people and resources and produces all the required outputs. Generally, though that is not a good idea for the following reasons:

- I may only need some of the people for some of the activities.
- I may not need to do all the activities all the time.
- I may need to look at other products beside the final outputs of a process.
- I may want to better manage a process by having different feedback and control at different times.

Therefore, people design process chains that link a set of processes together

The following is a simple example of a set of processes for creating software to meet a customer's needs.

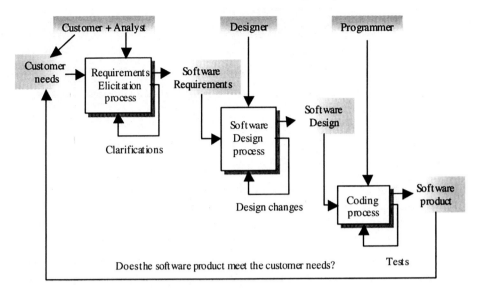

Simple software engineering process chain

This software engineering process chain produces the desired product (software) as well as several intermediate products (software requirements, software design). Different people may be involved in different processes (Customer, Analyst, Designer, Programmer) at different times, but they could all be the same person performing all roles (or a team that shares all the roles).

Whether performance is by one person or a team, the people involved still have a vital quality question to resolve:

"Does the software product meet the customer needs?"

If the answer is positive, then it can be said that the above process chain meets its purpose and the desired product outcomes have been achieved.

But is this the only purpose of the process chain?

Here are some other questions to ponder:

- Did I produce the software by the time that the customer wanted it?
- Did I produce the software at a cost that allowed me to make a profit (or at least break even)?
- Did the people involved work efficiently, or were some of them doing nothing some of the time?

- Can I follow the same process again to make another software product for another customer?
- Do I know how well all the activities were performed?
- Could I improve the way I do it next time to be faster and make it cost less?

A good way to answer these questions is by doing process assessment.

1.3 Concepts of Process Assessment

The previous section highlights what a process and a process chain can achieve (its purpose and outcomes) to meet the goals of the organization. It poses questions that are related to efficiency and effectiveness of the process chain itself, and the organization implementing these processes.

An organization can use various means to assess its performance. From a business viewpoint, it is important to consider why process assessment is a good means to do this.

To be efficient and effective, organizations need to know how well their current processes help them achieve their goals (amongst other factors). To be competitive, organizations also need to compare their efficiency and effectiveness with their competitors. Competitive forces will often mean that an organization needs to improve various aspects of their activities, including improvement of processes in order to remain competitive.

In addition, if the organization can relate what they used (inputs, resources) to the results they obtain, they can determine their efficiency. If an organization can compare their efficiency and effectiveness (achievement of their goals) to other organizations, they can judge their competitiveness.

When people assess processes, they increase their understanding of their actual performance and management of activities, and the potential for improvement.

Following the People-Process-Product model, people performing process assessment are assessing one of the 3 major factors leading to organizational success.

Simply stated, process assessment provides an organization with an important means to assess and improve their performance.

What specific aspects of process assessment are important?

In order to effectively and efficiently assess processes, the organization must use a clear and consistent measurement scale. The measurement scale should make clear distinctions between the levels of achievement. These clear levels of achievement not only help the organization understand its current performance, but can also act as guidance to what it should try to achieve in the future (improvement).

A simple analogy of this measurement scale might be to compare it to a home:

- Level 0 is like living outdoors on the bare earth under a tree. Not good.
- Level 1 is having a floor, four walls and a roof. Just basic shelter.
- Level 2 means you have doors and windows. Now you have some limited control of the environment inside your home (you can stop the wind and rain blowing in).
- Level 3 means you have basic services such as electricity and plumbing. Now you can better control the environment inside your home.
- Level 4 means that you have all the basic furniture and fittings. You are starting to feel at home and can think about what comforts you want.
- Level 5 means that you have all the desired modern conveniences, perhaps air conditioning, security, the latest home entertainment system, fast internet connection and so on. You are able to optimise you living environment to suit your lifestyle.

Naturally, a real process assessment measurement framework is a bit more complicated than the above analogy, but the principle of clear levels of achievement applies.

Furthermore, if the measurement scale is standardized, organizations that perform process assessment have one indirect form of comparison to competitors (their process capability).

In addition to a good measurement scale, process assessment requires a consistent approach (in fact a well defined process).

The following diagram illustrates some of the major activities and documented components that need to be addressed to ensure consistent process assessment.

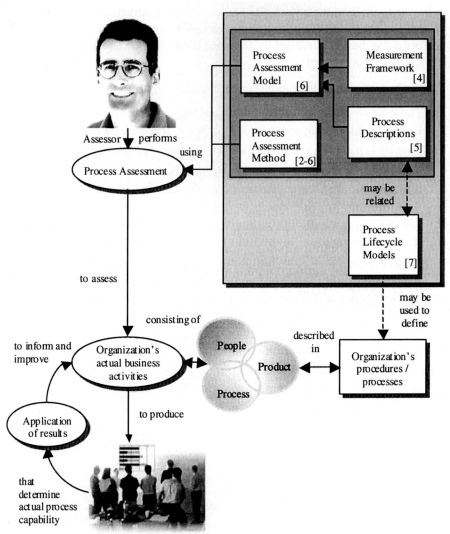

Process assessment overview.

As illustrated in the diagram, process assessment assesses the organization's actual business activities in terms of one or more implemented processes (process chains) against a (standard) measurement framework.

The characteristics of a process assessment approach that make it useful for an organization can be summarized as:

- A framework to ensure consistent assessments.
- A measurement framework that specifies a scale of performance (capability).
- A description of the processes the organization wishes to assess.
- A method to ensure consistent assessment performance by assessors.
- Guidance on how to tailor the assessment to suit the organization.
- Requirements on assessors to ensure they are capable and consistent in performing assessments.
- Guidance on how to apply assessment results.

These characteristics of process assessment are expanded in the following chapters of the book. To summarize these characteristics, we require that process assessment cover the following aspects.

✓ Process assessment compares the actual, implemented processes against a defined standard set of process characteristics (process attributes) consisting of process descriptions. The process descriptions must be clearly defined in terms of process purpose and outcomes [Chapter 5].

✓ Process assessment looks at how well a process is managed, by comparing the performance, management, definition, measurement and improvement of the process against a standard scale of good management practice. The measurement framework defines this scale for the measurement of process capability[2] [Chapter 4].

✓ Process assessment requires a way to apply the measurement framework and the process descriptions. This is defined in a Process Assessment Model [Chapter 6], which provides the extra level of detail required to ensure a repeatable process assessment result.

✓ Process assessment requires a consistent way to use the Process Assessment Model in an assessment. This is defined in a process assessment method [Practical Guide – Chapter 6].

✓ Process assessment is most effective when there is a way of adapting the assessment to suit the organization. This is especially important when assessment is used for improvement purposes. This requires guidance on tailoring and/or mapping of the chosen assessment model and method [Chapter 3].

---

[2] Process capability relates to the organization's ability to deliver specified or desired performance consistently/predictably.

✓ Consistent process assessment also requires a level of skill, experience and competence of the assessors [Chapter 8].

✓ A comprehensive process assessment approach will also provide guidance on interpretation and application of the assessment results (see the Practical Guide).

When an assessor performs process assessment, he/she will assess whether the organization has adopted some form of consistent approach for its processes. When an organization standardizes not just one or two processes, but defines a complete set of processes (process chains) so their interactions are also defined, they have a coherent process framework that improves their ability to succeed.

To help organizations decide how to define such a process framework, they can look at standards that define such frameworks in Process Reference Model standards [Chapter 5]. They can also look at more detailed implementation guidance contained in process lifecycle models [Chapter 7].

1.4 Process Assessment and ISO/IEC 15504

Process assessment can use a variety of approaches to achieve all of the requirements and needs in the previous section. Different models have been developed over time, many of which have been incompatible with each other.

ISO/IEC 15504 is a standard that provides a way to achieve all of the above. It specifies and/or refers to all the process assessment requirements needed to ensure consistent process assessments. ISO/IEC 15504 consists of 5 parts; part 2 is the normative part of the standard.

Many organizations have embraced the standard. They range from the international and national standards setting authorities, through professional organizations such as the Software Engineering Institute in the USA, the European Software Institute and the Software Quality Institute in Australia. They include large multinational and global firms and organizations such as the European Space Agency, France Telecom and automotive manufacturers, and even small and medium enterprises operating within a single country, state or city. The standard provides an ability to tailor it to suit each type and size of organization.

Organizations use the standard as a means to:

• assess the capability of their suppliers;

- provide a means to assure that consistent processes are applied in a manner to produce consistent quality products; and
- help their suppliers progress to better process management.

Most importantly, many organizations use ISO/IEC 15504 to improve their own business and this is the primary reason that use of the standard is so attractive.

One of the strengths of an ISO/IEC 15504 process assessment is that it defines where you are now and provides direction on how to improve your processes (capability). Improving process capability can help an organization's business aim to improve efficiency and effectiveness.

ISO/IEC 15504 in part 2 provides a complete framework for process assessment. The diagram below illustrates all the basic requirements for process assessment. It describes the four documented components required in the previous section, namely:

- The process assessment model.
- The measurement framework.
- The process descriptions.
- The process assessment method.

The latest revision of ISO/IEC 155504 part 2 specifies requirements on the externally specified process descriptions and process assessment methods.

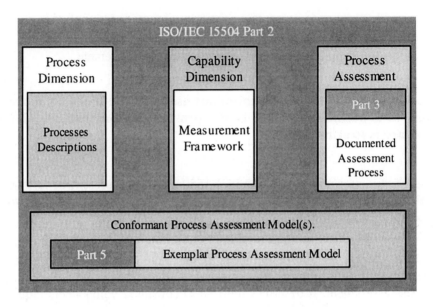

ISO/IEC 15504 process assessment components.

ISO/IEC 15504 Part 2 is aided by Part 3, which provides informative guidelines for application of the process assessment process (methodology), and by part 5 that provides an exemplar Process Assessment Model meeting the part 2 requirements for a conformant Process Assessment Model.

As shown in the above diagram, ISO/IEC 15504-2 specifies the requirements for process assessment using two complementary orthogonal dimensions:

- Capability dimension.
- Process dimension.

The two dimensions work together to provide complete process assessment ability.

The standard specifies the capability dimension as consisting of 6 Capability Levels (CL 0 – CL 5).

The standard specifies requirements for the process dimension consisting of a set of process descriptions (P1, ... Pn), which are defined in a Process Reference Model.[3]

---

[3] These processes are normally structured into process categories of related processes but for assessment purposes can also be considered as separate processes.

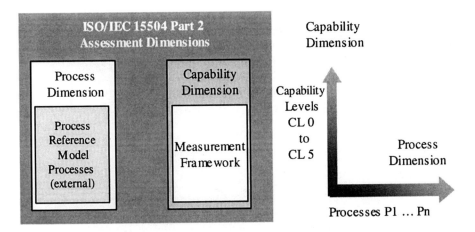

Two process assessment dimensions and the two dimensional continuous model.

ISO/IEC 15504-2 uses a continuous model representation[4] for the capability and process dimensions. The advantage of this two dimensional continuous model is that it allows for any process to be assessed and rated at any Capability Level, independently of any other process[5].

Therefore, the organization can decide to assess one process or a group of processes. It can select different groups of processes depending upon their importance to specific projects or organizational units (for example a customer help desk uses different processes to a project developing new software). The organization can also decide to which Capability Level it wishes to assess each process, for example the organization may decide to assess all processes to the same Capability Level, or it may decide to assess some processes to a higher Capability Level than others. This level of flexibility is very useful, as the organization may need some processes to be implemented with higher process capability in some projects/units due to their importance (for example in a safety critical development project).

The two dimensional model has other advantages compared to a staged model.

---

[4] Note: earlier process assessment methods consisted of a staged model (for example, Software Engineering Institute SW CMM® V1.1) where processes were required to be implemented in a specific order to indicate increasing process maturity. This is in some ways easier to follow, but much less flexible than using a Continuous Model (see the chapter comparing SW CMM® and CMMI® to ISO/IEC 15504 for further discussion of advantages and disadvantages).

[5] We will see later in the book that there are some logical, associated process dependencies.

It allows further orthogonal dimensions to be added. The following examples are possible three-dimensional models that illustrate the power of using orthogonal dimensions in the Continuous Model.

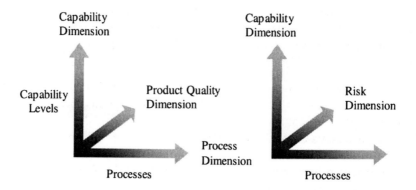

Three-dimensional models incorporating product quality or risk dimensions.

The ability to have additional dimensions allows the organization to compare these additional aspects to the process aspects. For example, if the organization has higher capability processes, how do they affect product quality and risk? For an example of use of the risk dimension, readers should look at the Practical Guide: Process Improvement chapter.

Further advantages of the Continuous Model are:

- The capability dimension is generally applicable to any process.
- An ability to use a variety of Process Reference Models without affecting the Capability dimension.
- An ability to use a variety of Process Assessment Models (and methods) for each Process Reference Model.
- An ability to use the Capability dimension with multiple International Standards (ISO 9000, ISO 12207, ISO 15288).
- There is a clear definition of processes (no mix of process description and how well the process is performed and managed).
- There is a clear definition of the Capability Levels and attributes of process capability.

Part 2 of the standard specifies the requirements for a Process Reference Model if the organization wishes to have an ISO/IEC 15504 conformant process assessment.

With the latest revision of part 2, this model is now specified externally to the standard. This allows an organization to choose and/or design their own Process Reference Model. The organizations may choose an existing model that suits their business (for example an existing software process standard or system engineering process standard), add processes that are unique to their way of doing business, or even create a complete process model. This flexibility also provides organizations that have used older standards a form of backwards compatibility that enables them to migrate to the new standard.

Also, part 2 specifies how an organization can validate the compliance of a conformant Process Assessment Model. One such conformant model is provided in part 5 of the standard as an exemplar. This model in part 5 meets all the requirements of part 2 for a Process Assessment Model for assessing an organization involved in acquiring, developing and/or supporting software oriented products and services. Furthermore, the model in part 5 is based on and compatible with ISO/IEC 12207[6].

The capability dimension has a measurement framework with nine process attributes that are grouped into Capability Levels 1 to 5. These Capability Levels (CL 1 – CL 5) define an ordinal scale of increasing process capability[7]. Note that Capability Level 0 is defined as an incomplete process and indicates a lack of achievement of the Capability Level 1 process attribute.

One of the strengths of this ordinal scale is that there is a clear level of achievement at each Capability Level. At Capability Level 1, the level of achievement is related to the performance of the process.

From Capability Level 2 to Capability Level 5, the level of achievement is related to how well the process is managed, defined, quantitatively controlled, and improved using a standard scale of good management practice. The Capability Levels apply consistently to all processes. This ordinal scale is a differentiating feature compared to most other assessment measurement frameworks.

---

[6] Part 5 is currently conformant to an earlier version of the standard (ISO/IEC TR 15504) but being revised.

[7] Another simple analogy is the Star rating of hotels (1 Star to 5 Star hotels). Each hotel offers a room with a bed, but as a hotel receives more stars, the quality and reliability of the accommodation improves.

	Capability Level	Process Attribute	Rating Scale
ISO/IEC 15504 Part 2	CL 0	None	NPLF
Capability Dimension	CL 1	PA 1.1	NPLF
	CL 2	PA 2.1, PA 2.2	NPLF
Measurement Framework	CL 3	PA 3.1, PA 3.2	NPLF
	CL 4	PA 4.1, PA 4.2	NPLF
	CL 5	PA 5.1, PA 5.2	NPLF

Rating Scale: N=Not, P=Partially, L=Largely, F=Fully achieved

Measurement Framework

In the following chapters, I will look at how the above dimensions are further developed into a Process Capability Measurement Framework, a Process Reference Model, and a Process Assessment Model.

In Annex 1, I briefly outline the history of the development of process assessment, including ISO/IEC 15504 and also describe one of the other objectives of the standard, which is to act as a means for harmonization for a wide variety of process assessment methods and models, so that assessment results can be exchanged.

2   ISO/IEC 15504 DOCUMENT SET

In this chapter, I describe the structure, relationships and assessment dimensions of the new standard as guidance to readers who need to understand the relationship between the various parts of the standard.

Assessors, quality professionals and other personnel involved in process design, implementation and assessment need to be aware of the process assessment dimensions and measurement framework so that processes are designed and assessed in a manner consistent with the requirements of the standard.

There is a reader guide to the standard for various roles.

ISO/IEC 15504 [1] consists of five parts as shown below. Part 2 is the normative part of the standard, and parts 1, 3, 4 and 5 are informative only (but provide important guidance and clarification).

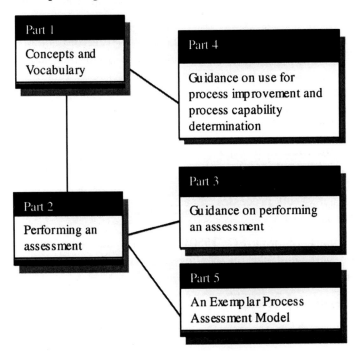

ISO/IEC 15504 Documentation.

The status of the document set of the standard at the time of authoring this book is as follows:

- ISO/IEC 15504-1 Information Technology — Process Assessment - Part 1 - Concepts and vocabulary: Committee Draft ballot completed.
- ISO/IEC 15504-2 Information Technology — Process Assessment - Part 2 - Performing an assessment: published.
- ISO/IEC 15504-3 Information Technology — Process Assessment - Part 3 - Guidance on performing an assessment: published.
- ISO/IEC 15504-4 Information Technology — Process Assessment - Part 4 - Guidance on use for process improvement and process capability determination: registered for FDIS ballot.
- ISO/IEC 15504-5 Information Technology — Process Assessment - Part 5 - An exemplar Process Assessment Model: Committee Draft circulated for review[8].

Part 1 of ISO/IEC 15504 provides a general entry point to readers of the standard. It describes how the parts of the document suite fit together, and provides guidance for their selection and use.

Part 2 is the normative part of ISO/IEC 15504 (i.e. the actual standard). ISO/IEC 15504-2 defines the 2-dimensional assessment model; consisting of a process dimension and a capability dimension. The process dimension refers to an external Process Reference Model (e.g. ISO/IEC 12207). A compliant model must provide a statement of process purpose and process outcomes. Part 2 sets requirements for verification of model compliance.

The capability dimension in part 2 has nine process attributes. The process attributes are grouped into five process Capability Levels (there are none at Capability Level 0), defining an ordinal scale of capability, which is applicable across all processes. It also sets requirements for performing conformant assessments and assessment models.

The third part of ISO/IEC 15504 provides an overview of process assessment with guidance on meeting the requirements for performing an assessment, including:

- An assessment process.

---

[8] Since the latest revision of ISO/IEC 15504-5 is only in Committee Draft form, when I describe the detailed aspects of this part later in the book, I will use the earlier technical report version: ISO/IEC TR 15504-5:1998. This does not affect the structure and relationships as described in this section.

- Guidance on the measurement framework for the process capability dimension.
- The requirements on the Process Reference Model and on process assessment methods.
- Guidance on competency of assessors

Readers with specific interest in either process improvement or supplier capability determination should read Part 4 for detailed guidance on these contexts of use. The concepts and principles of process improvement are generic and applicable to any type of organization or business. The guidance for process capability determination can be applied to any customer-supplier relationship. This part will enable the user to identify the appropriate usage of the normative components of ISO/IEC 15504 (part 2).

The fifth part of ISO/IEC 15504 provides an exemplar model compatible with the reference model in ISO/IEC 12207 AMD 1 [2]. This extends the reference model required in part two with a comprehensive set of indicators of process performance and capability for organizations that acquire, develop and operate software products.

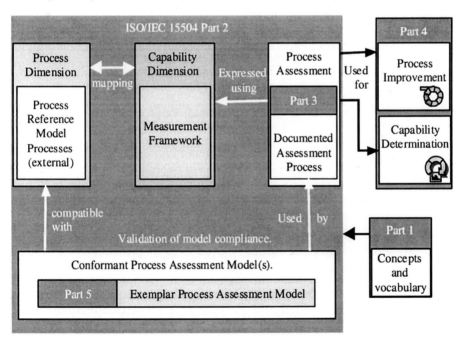

ISO/IEC 15504 relationship of parts

The diagram illustrates that the standard not only provides all the requirements for process assessment, but also provides guidance on how to use the results of process assessment (part 4), and requirements on validation of Process Assessment Models.

As various stakeholders in a process assessment need to use the standard, it is possible to guide readers. Table 1 identifies the principal readers for ISO/IEC 15504 and shows where their primary areas of interest are addressed within the document set.

Readership of parts of ISO/IEC 15504

Reader	Reader's Interests	Suggested parts
Assessment Sponsor	How an assessment is conducted, what tools and other support are required, how to initiate an assessment.	1, 2, 3
Process Improvement Sponsor	Initiating an improvement programme, defining assessment inputs for an assessment for improvement purposes, using assessment results for improvement.	4
Process Capability Determination Sponsor	Initiating a programme for the determination of supplier capability, defining a target capability profile, verifying and using assessment results in a capability determination exercise.	4
Assessors	Conducting a conformant assessment, developing the skills and competencies needed to perform an assessment.	2, 3, 4, 5
Developers of Process Assessment Models	Developing Process Assessment Models for performing assessments based on a compliant Process Reference Model and the measurement framework as defined in ISO/IEC 15504-2	2, 3, 5
Developers of Assessment Methods	Developing a method that will support the performance of conformant assessments.	2, 3, 5
Tool Developers	Developing tools that will support assessors by collecting, recording and classifying evidence in the performance of assessments.	2, 3, 5

As the documentation set is quite large, it is inappropriate (or too time consuming) for everyone in the organization to read all parts of the standard. Therefore, the reading guide should be used to guide persons having various roles within the organization being assessed.

If an organization is preparing an assessment, the persons (normally the assessors) involved should consider creating a presentation that covers the main aspects of the assessment process and the improvement process for the sponsors, rather than relying on the sponsors to read the relevant sections of the standard.

3   MEETING THE REQUIREMENTS OF ISO/IEC 15504

ISO/IEC 15504 has a requirement to harmonize various Process Assessment Models and methods to ensure that assessment results are compatible and can be used in a variety of organizations. This affects the various assessment models, assessment methods and certification aspects and in this chapter, I describe how the harmonization requirements set explicit requirements for conformance for these aspects.

This chapter will be of interest to assessment model developers, assessors and assessment sponsors.

The ISO/IEC 15504-2 requirements for conformance specify relationships between a Process Reference Model, a Process Assessment Model and a Conformant Assessment process and are shown in the figure below.

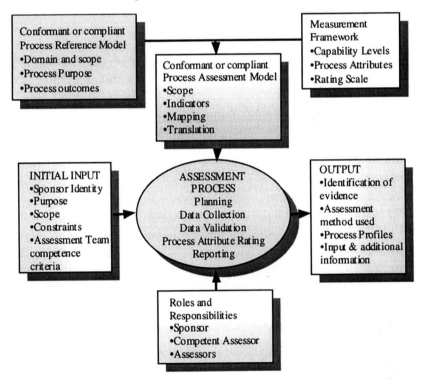

Process Reference Model, Assessment Model and Assessments relationships

ISO/IEC 15504-2 specifies the Process Capability measurement framework.

ISO/IEC 15504-2 also specifies the need for a Process Reference Model (which specifies the processes to be assessed).

The Process Reference Model provides the mechanism whereby a Process Assessment Model(s) is related to the measurement framework defined by ISO/IEC 15504-2.

The Process Reference Model is defined external to ISO/IEC 15504-2 (see the Process Reference Model chapter for examples), so that it can be changed to suit the organization.

The Process Assessment Model is based on the process descriptions provided in Process Reference Models (see the Process Assessment Model chapter for examples).

Finally, ISO/IEC 15504-2 also specifies the requirements for a conformant assessment process using the selected Process Assessment Model.

3.1   Conformant Process Reference Model

An ISO/IEC 15504 conformant Process Reference Model shall adhere to certain requirements to assure that assessment results are translatable into an ISO/IEC 15504 process profile in a repeatable and reliable manner. The model shall fulfil the following requirements:

- A declaration of the domain of application (for example, software development, space software, automotive).
- Description of the processes[9] shall include:
  - Purpose and outcomes.
  - The set of process outcomes shall be necessary and sufficient to achieve the process purpose. This could be production of an artefact, a significant change of state or meeting specified constraints such as requirements or goals.
  - Unique process descriptions and identification for each process.
- Process descriptions shall not contain or imply aspects of the measurement framework beyond Capability Level 1.
- A description of the relationship between the Process Reference Model and its intended context of use within its domain of application.

---

[9] See clause 6.2.4 of ISO/IEC 15504-2

- A description of the relationship between the processes defined within the Process Reference Model.
- The Process Reference Model shall document the community of interest of the model. In addition, it shall describe the actions taken to achieve consensus within that community of interest.
  - The relevant community of interest will be characterized or specified.
  - The extent of achievement of consensus shall be documented, or if no consensus actions are taken, a statement to this effect.
- Verification of the extent to which the model meets the requirements of ISO/IEC 15504 may be through either demonstration of conformity or demonstration of compliance[10].

3.2 Conformant Process Assessment Model

ISO/IEC 15504-2 specifies that a conformant Process Assessment Model shall be based upon a suitable reference source of process definitions, which is called a Process Reference Model.

It is feasible to have different Process Assessment Models using the same Process Reference Model basis, as long as they meet the conformance requirements through their relationship with the specified Process Reference Model.

It is also feasible to have one Process Assessment Model using several Process Reference Models for its basis, but this is more difficult to achieve (and the author has yet to see one in practice). However, one Process Assessment Model has industry specific extensions in addition to a standard Process Reference Model[11].

A conformant Process Assessment Model provides a two-dimensional view of process capability[12].

In order to ensure consistency and repeatability of assessments, a conformant Process Assessment Model shall contain:

---

[10] The party performing verification shall obtain objective evidence that the Process Reference Model fulfils the requirements set forth in ISO/IEC 15504-2 clause 6.2. Objective evidence of conformance or compliance shall be retained.

[11] SPiCE for SPACE used ISO/IEC TR 15504-2:1998 Process Reference Model and added space industry specific extensions. The latest revision will use the ISO/IEC 12207 Process Reference Model.

[12] The model must provide at least a process and a capability dimension, it may optionally provide more.

- A definition of its purpose, scope, elements and indicators.
- A mechanism for consistent expression of results.
- The mapping to the ISO/IEC 15504 Measurement Framework and the specified Process Reference Model(s).

The Process Assessment Model shall be based on process management principles and designed for assessing process capability, addressing all of the Capability Levels of the measurement framework for each of the processes within its scope. Note: the standard allows for the possibility that models may only address a continuous subset of the Capability Levels starting at level 1 (but model developers are unlikely to use this option).

The Process Assessment Model shall declare its scope of coverage in the terms of:

- The selected Process Reference Model (s);
- The selected processes from the Process Reference Model (s);
- The Capability Levels selected from the measurement framework.

The conformant Process Assessment Model shall use a set of process capability indicators (PCI). The process capability indicators shall explicitly address the process purpose(s) of the selected processes from the selected Process Reference Model(s). They will demonstrate the achievement of the process attributes within the selected Capability Levels.

The Process Assessment Model shall provide a formal and verifiable rating mechanism for representing the assessment results as a set of process attribute ratings for each (assessed) process. The representation of results may involve a direct translation of process attribute ratings into a process profile[13] as defined in ISO/IEC 15504. Alternately, it may use the conversion of the data collected during the assessment (and any additional information collected) through further judgment on the part of the assessor.

Conformant Process Assessment Models shall enable comparison of outputs from assessments based upon the same Process Reference Model, using different Process Assessment Models.

---

[13] Process profile = set of process attribute ratings for a process

3.3 Mapping Process Assessment Models to Process Reference Models

ISO/IEC 15504-2 specifies the need for mapping between the various models.

A Process Assessment Model shall provide an explicit mapping from the relevant elements of the model to the processes of the selected Process Reference Model and to the relevant process attributes of the measurement framework.

The mapping shall be complete, clear and unambiguous. The mapping of the indicators within the Process Assessment Model shall be:

a) the purposes and outcomes of the processes in the specified Process Reference Model;

b) the process attributes (including all of the results of achievements listed for each process attribute) in the measurement framework.

This enables Process Assessment Models that are structurally different to be related to the same Process Reference Model.

[ISO/IEC 15504-2, 6.3.4]

The mapping provides the way to translate ratings from an ISO/IEC 15504 conformant assessment into a common format, which makes it possible to compare assessment ratings (from different assessments and from different Process Assessment Models).

The mapping requirements in the standard specify that the Process Assessment Model be accompanied by a detailed set of mappings, which demonstrate:

- How the indicators of process performance provide coverage for the purposes and outcomes of the processes in the specified Process Reference Model.
- How the indicators of process capability within the model provide coverage for the process attributes (including all of the results of achievement of the process attributes) in the measurement framework.

In the following diagram, the mapping requirements for the process capability are shown.

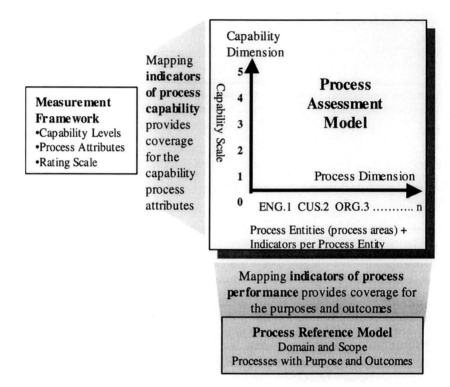

Mapping Process Assessment Model, Process Reference Model and Capability.

For an assessment to be considered conformant, it is essential that the assessor has access to the details of the mapping of the elements of the model to the Process Reference Model.

The mapping may be straightforward, for example, the model defined in technical report version of the standard (ISO/IEC TR 15504-5) uses the processes of the Process Reference Model directly as the basis for its definition of the processes (one to one process mapping)[14]. In addition, the ISO/IEC TR 15504-5 Process Assessment Model employs a continuous model architecture (one to one capability dimension mapping).

---

[14] The Process Reference Model was contained in the earlier ISO/IEC TR 15504-2:1998 version of the standard, but has been removed in the issued version of the standard ISO/IEC 15504-2:2003. It is based upon ISO/IEC 12207.

Where the structure of the model is significantly different from the Process Reference Model, as could be the case of a Process Assessment Model employing other architectures[15], the mapping will likely be more complex.

In the early development of the standard (ISO/IEC PDTR 15504-5, the preliminary draft technical report version of the standard), the assessment trials indicated several instances where elements were components of more than one process attribute, This required modification to the model which was incorporated in the draft technical report and in the technical report version: ISO/IEC TR 15504-5.

Mappings that result in elements being identified as components of more than one process attribute may indicate problems with the model structure, which could result in ambiguous translation of results.

Therefore, a Process Assessment Model developer should ensure that the mapping is clear and unambiguous. The standard requires that the model developer provide proof that the model conforms to the requirements for mapping clarity and translation of results.

An assessor should confirm that the mapping is meaningful – especially when trying a new model or when the model developers indicate that the model is not yet fully proven. One way to confirm that the mapping is meaningful is to sample some of the lowest level components in the model, and locate them in the Process Reference Model, either as elements of a process or as contributors to a process attribute.

3.4   Conformant Assessments

ISO/IEC 15504 defines the requirements for a conformant assessment. The requirements can be summarized as follows:

- Use a documented assessment process, which must be capable of meeting the assessment purpose[16].
- Define roles and responsibilities of the sponsor, the competent assessor and assessors.
- Define the assessment inputs for each assessment.
- Conduct the assessment according to the documented assessment process.

---

15 An example is the staged representation of the SEI CMMI® .

16 ISO/IEC 15504 defines assessment purpose as "a statement, provided as part of the assessment input, which defines the reason for performing the assessment."

- Record the assessment output according to the standard.

The documented assessment process shall contain at minimum the following activities:

a) **Planning** — a plan for the assessment shall be developed and documented, including at minimum:

   1) the required inputs specified in this part of ISO/IEC 15504;

   2) the activities to be performed in conducting the assessment;

   3) the resources and schedule assigned to these activities;

   4) the identity and defined responsibilities of the participants in the assessment;

   5) the criteria to verify that the requirements of this International Standard have been met;

   6) a description of the planned assessment outputs.

b) **Data collection** — data required for evaluating the processes within the scope of the assessment (see 4.4.2 c)) and additional information (see 4.4.2 j)) shall be collected in a systematic manner, applying at minimum the following:

   1) the strategy and techniques for the selection, collection, analysis of data and justification of the ratings shall be explicitly identified and shall be demonstrable;

   2) correspondence shall be established between the organizational unit's processes, specified in the assessment scope, and the elements in the Process Assessment Model;

   3) each process identified in the assessment scope shall be assessed on the basis of objective evidence;

   4) the objective evidence gathered for each attribute for each process assessed shall be sufficient to meet the assessment purpose and scope;

   5) the identification of the objective evidence gathered shall be recorded and maintained to provide the basis for verification of the ratings.

c) **Data validation** — the data collected shall be validated to:

   1) confirm that the evidence collected is objective;

   2) ensure that the objective evidence is sufficient and representative to cover the scope and purpose of the assessment;

   3) ensure that the data as a whole is consistent.

d) **Process attribute rating** — a rating shall be assigned based on validated data for each process attribute:

   1) the set of process attribute ratings shall be recorded as the process profile for the defined organizational unit;

2) during the assessment, the defined set of assessment indicators in the Process Assessment Model shall be used to support the assessors' judgement in rating process attributes in order to provide the basis for repeatability across assessments;

3) the decision-making process that is used to derive rating judgements shall be recorded;

4) traceability shall be maintained between an attribute rating and the objective evidence used in determining that rating;

5) for each process attribute rated, the relationship between the indicators and the objective evidence shall be recorded.

e) **Reporting** — the assessment results, including at minimum the outputs specified in 4.5, shall be documented and reported to the assessment sponsor or to their delegated representative.

[ISO/IEC 15504-2, 4.2.2]

Some clarification of the requirements of the standard is provided in the following text, including guidance on practical aspects associated with process assessments.

Planning

Planning depends in the first place on the assessment purpose (e.g. capability determination or improvement) and scope. The purpose and scope influence the processes to be assessed, the activities to be performed, the resources required (both people, places and tools) and the schedule proposed. The scope also needs to specify the parts of the organization, i.e. the organizational unit(s), to be assessed. An organizational unit may be a project team, a department, a business unit or some other form of organizational entity.

It is preferable that the assessors work with the sponsor and the organizational unit(s) representative when planning the assessment(s).

An assessment consumes organizational resources. Therefore, it is important that there are clearly defined roles and responsibilities for the main assessment participants in order to help maximize the efficiency and effectiveness of the assessment process. As most assessment methods rely substantially on interviews with people executing the process, it is important to carefully select who to interview, and arrange the interview times and facilities in the planning stage. People interviewed should be able to describe the processes and access/provide needed data to illustrate the process implementation.

In addition, during planning, the assessment inputs and outputs need to be stated and agreed. These may include project documents, organizational unit process descriptions and process implementation data, and outputs may include preliminary and final reports, planned discussion sessions on the results with participants, the sponsor and management.

The planning should also refer to the Process Assessment Model being used, any mapping between the organization's processes and those of the model, and any tailoring or criteria applied that demonstrate compliance to the requirements of ISO/IEC 15504.

Data Collection

During the assessment, the assessors need to collect evidence to support the assessment scope and purpose. The assessors need to collect data for each process assessed. This may be collected in a variety of ways including interviews, questionnaires, general discussions and data/document review. The strategy and techniques should support the assessment purpose and be explicitly documented.

The organizational unit may need to illustrate how they have implemented the process and whether it is an existing standard process or a modified or unique process implementation.

In the case of a modified implementation, the assessor should try to determine how the information provided by the organizational unit illustrates this modified/tailored process implementation. This explicit step should also check the relationship to the Process Assessment Model. It is preferable that any such checking occurs before the assessment, but assessors may often only become aware of this during the data collection. In some cases, there may be an explicit mapping of the implemented (defined) process.

There may also be iteration of which processes are sampled (for example, the same process may be sampled in several projects). The decisions on selection and sampling should be noted to guide other personnel (and can be useful for repeating assessments later).

The data may be manually or automatically collected and reported (e.g. project dashboard type data collection at defined project milestones). In both cases, the evidence collected needs to be retained to allow data validation and later verification of the assessment ratings.

Data Validation

This activity in the assessment is concerned with checking that the data collected is representative of the processes assessed, both in terms of sampling (organizational units, data types and processes) and in meeting the assessment purpose and scope. Furthermore, the assessors need to check that the data collected across processes (especially related or connected processes) is consistent (for example, document and configuration management).

In cases where the assessment is a repeat assessment used to confirm process improvements, the assessor may compare the data against that collected in prior assessment to confirm the improvements.

The assessors should use the preliminary results presentation to the organization as another basis for validating that the data represents the organizational unit assessed. When the preliminary results are in dispute, the assessors need to collect new or more data to validate their findings. If this proves impossible, then the assessors need to clearly state this in the assessment report together with an analysis of what this means in terms of rating, improvement opportunities and risks of misinterpretation of the results.

Process attribute rating

The assessors need to use their expertise when performing the process attribute rating. The rating is based upon the evidence (data) collected as much as possible, measured against the assessment indicator(s), but may still require some expert judgment and interpretation by the assessors.

The judgment should explicitly refer to sufficient evidence so that other organizational personnel (or subsequent assessments) are able to conclude that the ratings are valid for each process attribute assessed. In the simplest case, the rating components should accurately reflect the Process Assessment Model, the Process Reference Model (i.e. process purpose) and have a clear traceability to ISO/IEC 15504 process attributes.

Normally the assessors should agree on the rating (or use some documented rule in the case that agreement is not unanimous).

Note: In the author's experience, when assessments are performed for internal improvement purposes, it is preferable to also involve the organizational unit personnel in the rating. The reason to do this is that the personnel are more likely to adopt a positive attitude to the assessment and subsequently be more willing to perform improvements.

The set of process attribute ratings constitutes the process profile for the assessed organizational unit. This process profile should be presented in a manner that allows straightforward interpretation of meaning and value.

Reporting

Assessment reporting can focus on two types of detail:

- Capability determination and associated risks of capability target gaps.
- Improvement opportunities.

Normally the assessors should make a preliminary presentation of the results as part of the data validation to ensure that their final report reflects the assessed organizational unit performance, any feedback from the unit, and provides a clear and easy way to interpret the result. The final reporting should also reflect the assessment purpose (for example, improvement and/or capability determination focus), together with agreed supplementary information (for example, improvement opportunities, action plans, benchmarks, comparison to prior assessment results or against other organizational units – see the Practical Guide for templates).

Roles and responsibilities of involved personnel

The roles and responsibilities for the key personnel in assessments must be defined, including:

- The assessment sponsor, who:
  - should verify that the assessment has a competent assessor with the necessary competence and skills;
  - ensures that adequate resources are available to conduct the assessment, including key personnel for interviews, infrastructure, and documents to be examined; and
  - ensures that the assessment team has access to the relevant resources, including the assessed organizational unit management as needed.
- The competent assessor, who
  - shall confirm the sponsor's commitment to proceed with the assessment;
  - ensures that the assessment meets the agreed purpose;
  - ensures (and verifies) that the assessment is conformant to ISO/IEC 15504;
  - ensures that participants are briefed on the purpose, scope and approach at the start or during the assessment (normally there should

be an introductory briefing for the main personnel, with shorter briefings for interviewed personnel who did not attend the introductory briefing);

- ensures that the assessment team members have appropriate knowledge and skills, including use of any assessment tools;
- ensures that the assessment team members have access to appropriate documented assessment activities guidance; and
- confirms that the sponsor receives the assessment result deliverables.

In addition, the role and responsibilities of assessors and assessment coordinators (for example the assessed organizational unit coordinator) should be clearly defined and documented.

The assessors shall:

- carry out assigned activities including planning, data collection, data validation and reporting; and
- rate the process attributes.

In summary, ISO/IEC 15504 requires the combination of a documented assessment process and competent personnel for a conformant assessment.

3.5   Opportunities for Certification

There are several areas in which certification is relevant. They include:

- Certification of Assessors.
- Certification of Assessment Results.
- Certification of conformance of a Process Reference Model.
- Certification of conformance of a Process Assessment Model.

In the standard, there are several mechanisms for verification of conformity to verify that the requirements of ISO/IEC 15504-2 have been fulfilled.

The conformity to the requirements of ISO/IEC 15504 may be verified by self-declaration (first party), by a second party, or by a third party. A first party could be a model developer or user. A second party could be a community of interest, for example, the group of automotive manufacturers involved in defining Automotive SPICE or the European Space industry for SPiCE for SPACE. A third party could be the national standards body, an independent certification body, or potentially ISO or the IEC.

In essence, if a Process Reference Model meets the requirements of clause 6.2 in ISO/IEC 15504-2:2003 it can be verified by those involved in its development and use, and submitted for conformity to the relevant community of interest. If a Process Assessment Model meets the requirements of clause 6.3 in ISO/IEC 15504-2:2003 it can be verified by those involved in its development and use and submitted for conformity to the relevant community of interest.

If a process assessment result meets the requirements of clause 4 in ISO/IEC 15504-2:2003 (see section 5.4 above) and it is verified by a competent assessor, it can be submitted to the community of interest. In general, the community of interest responsible for a particular assessment model will only certify that the assessment was conducted in conformance with the requirements and not certify the assessment result (for example the Software Engineering Institute will register assessment results for SW CMM® and CMMI® assessments, but not certify that the organization assessed is at a particular maturity Level).

It is important that readers recognize the difference between certification of an assessment result and certification of performance of an assessment.

In the former case, the organization could claim certification that it performs at an internationally recognized standard of process performance and this places a liability on the certifying authority. In the latter case, the liability is removed from the certifying authority and the assessment result becomes the responsibility (with any resultant liability) of the competent assessor.

Hence, it is common practice for the competent assessor to explicitly state that the assessment was properly performed and applies only to the organizational instance(s) assessed. For example, in Organization 1, for project XYZ the processes A, B, C were assessed at a particular Capability Level.

Therefore when reviewing organizational capability or maturity level claims, readers should not just note the claimed performance, but also the scope of the assessments performed (and by whom) in order to decide how well the claimed performance will reflect performance for their specific needs.

Certification of Assessors also requires achievement of particular ISO/IEC 15504 requirements, including the assessors:

* general level of education;
* experience;
* assessor training by a recognized training provider (e.g. SQI, SYNSPACE, Software Engineering Institute);

- performance of assessments under the guidance of a recognized competent assessor; and
- leading an assessment(s) under the guidance of a recognized competent assessor.

The chapter on **Assessor Competence** describes the guidance on competence of assessors in more detail.

4   THE MEASUREMENT FRAMEWORK

The measurement framework of the standard is focused on the capability dimension as described in chapter 3.

In this chapter, I describe the measurement framework at the overall Capability Level, then in more detail for each process attribute for each Capability Level.

I provide further information for each process attribute under the headings: clarification, state of practice experience, and guidance to users on changes to the standard.

The rating scale, known as NPLF (meaning Not, Partially, Largely, Fully achieved) is described. The next section in the chapter describes the way the process attributes map upwards to the Capability Levels.

Finally I make a comparison between ISO/IEC 15504 and ISO 9004:2000, which also has a (different) maturity dimension and provide a mapping of the process and capability dimensions derived in the ESA S9KS study project by SYNSPACE.

This chapter will be of interest to anyone designing or assessing processes, including assessors, process owners, process experts, and developers of assessment models.

4.1   Capability Levels ISO/IEC 15504

In part 2 of the standard, ISO/IEC 15504 defines a measurement framework for the assessment of process capability. The measurement framework provides a mechanism for evaluating the capability of an implemented process with respect to a Process Assessment Model.

The Capability dimension of ISO/IEC 15504 consists of 6 Capability Levels. This process capability is defined on a six point ordinal scale that enables capability to be assessed and expressed on a scale ranging from **Incomplete** at the bottom of the scale to **Optimising** at the top end of the scale.

At **Incomplete**, the basic purpose of the process is not achieved. At **Optimising**, the process meets current and projected (expected future) business goals. The scale therefore represents increasing process capability of the implemented process. It thereby also defines a possible route for improvement for each process.

In brief, the Capability levels are:

- Incomplete: The process is not implemented, or fails to achieve its process purpose.
- Performed: The implemented process achieves its process purpose.
- Managed: The process is now implemented in a managed fashion.
- Established: The process is now implemented using a defined process that is based upon a standard process and that is capable of achieving its process outcomes.
- Predictable: The process now operates within defined limits to achieve its process outcomes.
- Optimising: The process is continuously improved to meet relevant current and projected business goals.

The Capability Levels are dependent and cumulative. The higher Capability Levels are dependent upon meeting the lower Capability Levels first, and the assessment rating accumulates as each level is attained. This means that during an assessment the process capability must first meet the requirements of Capability Level 1 (CL 1), before it can be rated at Capability Level 2 (CL 2), and so on.

The Capability Levels are further divided into Process Attributes (and management indicators). These aid assessors to ensure that reliable and sufficient evidence is available that a process achieves (or does not achieve) a Capability Level.

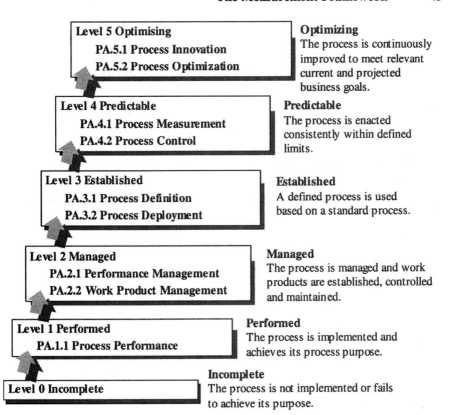

Level 5 Optimising	Optimizing
PA.5.1 Process Innovation	The process is continuously improved to meet relevant current and projected business goals.
PA.5.2 Process Optimization	

Level 4 Predictable	Predictable
PA.4.1 Process Measurement	The process is enacted consistently within defined limits.
PA.4.2 Process Control	

Level 3 Established	Established
PA.3.1 Process Definition	A defined process is used based on a standard process.
PA.3.2 Process Deployment	

Level 2 Managed	Managed
PA.2.1 Performance Management	The process is managed and work products are established, controlled and maintained.
PA.2.2 Work Product Management	

| Level 1 Performed | Performed |
| PA.1.1 Process Performance | The process is implemented and achieves its process purpose. |

| Level 0 Incomplete | Incomplete |
| | The process is not implemented or fails to achieve its purpose. |

Capability Dimension illustrating the Capability Levels and Process Attributes.

The Capability levels have been refined in successive versions in the development of ISO/IEC 15504. In brief, the latest version of the standard contains redrafting to remove ambiguity and increase consistency with ISO 9001:2000. This resulted in Capability Level three (CL 3) being restructured into definition and deployment. At Capability Level four (CL 4), the definitions have been clarified to make the concept of "quantitative understanding" clearer, and at Capability Level five (CL 5), there have been significant changes in the names and descriptions of the attributes of process capability, although the capability issues addressed substantially the same.

The latest version aids clarity and ensures consistent interpretation of the various Capability Levels.

4.2   Process Attributes

Within the measurement framework, the measure of capability is based upon a set of Process Attributes (PA). Process attributes are used to determine whether a process has reached a given capability. Each attribute measures a particular aspect of the process capability. The attributes are themselves measured on a defined scale and therefore provide a more detailed insight into the specific aspects of process capability required to support process improvement and capability determination.

The process attributes are defined in such a way that they can be rated independently of one another. However, the attainment of some attributes may be linked to the attainment of another attribute within the capability dimension.

The following description of the Capability Levels and process attributes (in boxes) is taken from ISO/IEC 15504 verbatim to ensure accurate definition.

Notes. Further information from the author is provided under the headings: clarification, state of practice experience, and guidance to users on changes to the standard.

The term 'organization' is used in this chapter generically; it does not imply a particular grouping of people, project team, department, business unit, an entire corporation/public body, or grouping of corporations/public bodies.

The term organizational unit (OU) defines the particular grouping of people, for example a project team, a department or a business unit that produces the desired products and results.

Level 0: Incomplete process

Level 0: Incomplete process

The process is not implemented, or fails to achieve its process purpose.

At this level there is little or no evidence of any systematic achievement of the process purpose.

[ISO/IEC 15504-2 5.1]

Clarification

Emphasis here is whether there is performance of a process to achieve its **purpose or not**. This Capability Level may appear to be superfluous, but in fact, the amount of activity may vary from 'not performed at all' to performed but does not achieve the complete process purpose. Secondly, it is possible that the process is performed, but so inconsistently or irregularly over time or in different organizational units (i.e. projects, sections, departments, business units) to be considered not systematic. There is also no systematic evidence (work products) that demonstrates that the process is performed. Work products are artefacts associated with the execution of a process.

An example could be testing; in some cases testing is performed, on other occasions due to schedule pressure, the testing is not performed or only partially performed with the result that faults are not found, which testing when performed would have found.

State of Practice Experience

In the author's experience organizations are sometimes rated at level 0 due to a failure to understand some fundamental aspect of the purpose of the process and hence implement an activity to achieve this aspect. For some technical disciplines in software development, there may be 6 to 10 fundamental activities to perform. This level also quickly illustrates to management the missing fundamentals in their organization's processes.

Level 1: Performed process

Level 1: Performed process

The implemented process achieves its process purpose.

The following attribute of the process demonstrates the achievement of this level:

PA 1.1 Process performance attribute

The process performance attribute is a measure of the extent to which the process purpose is achieved. As a result of full achievement of this attribute:

a) the process achieves its defined outcomes.

[ISO/IEC 15504-2 5.2]

Clarification

Firstly, the purpose of the process implemented within the organization is understood. This means that the organization can state the purpose of their process (the 'why') in terms of the outcomes it should achieve. This statement of purpose and the desired outcomes may be documented or may be the common view of the persons implementing the process.

Secondly, the organization routinely performs the fundamental activities needed to achieve the purpose and create the desired outcomes to a greater (or lesser) extent. This performance is demonstrated by some form of process outcome, which could be:

• Production of an artefact (document, product);

- A significant change of state (team understands product requirements); or
- Meeting a specified constraint (goal, requirement, etc.).

The organization's fundamental activities and the outcomes will naturally vary for each process.

Assessors will try to verify that the people performing the process understand the purpose of the process itself and perform the necessary activities. The assessors determine if these practices have been performed in a systematic and (fairly) consistent manner.

If necessary, the assessor may refer to the applicable Process Reference Model to determine the types of fundamental activities that are relevant to indicators called Base Practices in the Process Assessment Model. This will help determine whether these indicators show that the process outcomes are achieved.

The assessors will look for process performance indicators that meet the requirements of process attribute [PA1.1]. These may consist of the following types of indicators:

- identified work products that are input to the process; and/or
- identified work products that are produced by the process; and/or
- actions taken to transform the input work products into output products.

The work products resulting from performing the activities, together with input work products, are evidence of process performance when they contribute to achieving the process purpose. If they do not contribute to achieving the process purpose, they are not considered relevant process performance indicators. Note that processes performed at this level are not *assumed* to be meeting time, budget or other management restraints.

State of Practice Experience

In the author's experience, most organizations that understand the purpose of the process have also researched the required fundamental activities (base practices) and outcomes in order to achieve the process purpose.

Organizations that do not fully achieve level 1 generally miss one activity equivalent to a base practice or cannot show evidence of achievement.

This is especially true in organizations that have evolved their practices over time into new areas of application and have not accordingly adapted their processes.

More often, the organization produces artefacts based on templates that may not take into account the new application and thereby miss some important evidence of process performance. In most cases, the assessed organization can relatively simply complete the missing evidence.

A note to assessors: the indicators of process performance listed above are complementary – any one of them may be sufficient evidence of performance, it does not necessarily require all of them. Also, the Process Reference Models and Process Assessment Models will list many work products, only some of which may be relevant and needed by the organization and therefore be relevant indicators of process performance.

Guidance to users of earlier versions of ISO/IEC 15504 Technical Reports

The Capability Level remains basically unaltered from earlier versions of the standard, although the earlier technical report version explicitly required identifying the scope of work, which is assumed to be in the defined outcome achievement.

Level 2: Managed process

Level 2: Managed process

The previously described *Performed process* is now implemented in a managed fashion (planned, monitored and adjusted) and its work products are appropriately established, controlled and maintained.

The following attributes of the process, together with the previously defined attributes, demonstrate the achievement of this level.

PA 2.1 Performance management attribute

The performance management attribute is a measure of the extent to which the performance of the process is managed. As a result of full achievement of this attribute:

a) objectives for the performance of the process are identified;

b) performance of the process is planned and monitored;

c) performance of the process is adjusted to meet plans;

d) responsibilities and authorities for performing the process are defined, assigned and communicated;

e) resources and information necessary for performing the process are identified, made available, allocated and used;

f) interfaces between the involved parties are managed to ensure both effective communication and also clear assignment of responsibility.

PA 2.2 Work product management attribute

The work product management attribute is a measure of the extent to which the work products produced by the process are appropriately managed.

As a result of full achievement of this attribute:

a) requirements for the work products of the process are defined;

b) requirements for documentation and control of the work products are defined;

c) work products are appropriately identified, documented, and controlled;

d) work products are reviewed in accordance with planned arrangements and adjusted as necessary to meet requirements.

NOTE 1 Requirements for documentation and control of work products may include requirements for the identification of changes and revision status, approval and re-approval of work products, and for making relevant versions of applicable work products available at points of use.

NOTE 2 The work products referred to in this Clause are those that result from the achievement of the process outcomes.

[ISO/IEC 15504-2 5.3]

Clarification

At level 2, the organization manages the performance of the processes. This is the first Capability Level at which explicit management of the process occurs. This means that the performance of the process is planned, monitored and adjusted to meet organization needs and objectives.

The organization selects and uses work products that help them meet the objectives of the process and demonstrate achievement of the process outcomes. These work products are appropriately identified, documented and controlled.

The management of the process is focused on ensuring that the organization knows what to produce (work products), by when it needs to be produced (plan, schedule), that the outputs meets the organization needs, and that the process more reliably meets the planned performance.

The management of the process will result in artefacts and/or activities, which are verifiable (e.g. planning and/or plans, monitoring mechanisms, and/or process control mechanisms, and/or adjustments to the process based upon the results of comparison of the planned versus actual performance of the process.)

PA2.1 Performance Management attribute

The performance management attribute is concerned with the application of management techniques to provide reasonable assurance that process performance objectives are met.

The organization must first identify the process performance objectives, which could include one or more of the following:

- time planned and taken for the process (i.e. process cycle time);
- resources used in the process (e.g. people, products, finances, time);
- quality of the outputs/artefacts produced (e.g. performance, reliability, usability).

The organization also sets the objectives for process performance based on a set of internal and external considerations, including:

- organizational unit constraints (e.g. deadlines) and characteristics (e.g. research and development, maintenance projects);
- product constraints (e.g. cost) and characteristics (e.g. innovative technology, quality robustness);
- process inputs from another process or external source (e.g. client); and/or

- personnel constraints (e.g. availability of required personnel, training needs).

At level 2, the process performance objectives may be either qualitative terms (e.g. tests will be easy to understand and to conduct) or quantitative terms (e.g. tests will detect at least 90% of the defects in the product).

The organization's process performance objectives allows them to apply management techniques to provide reasonable assurance that process performance objectives are met.

For a process to be effectively performed, the organization must ensure that the personnel are empowered (have clear lines of authority) to perform the activities, and have a clear commitment to be responsible for the performance desired. Without authority, commitment and responsibility, the performance of the process is at risk of failure to achieve the purpose. Hence, the organization must ensure explicit assignment of responsibility and authority for performing the process. The essential aspects to be addressed are the identification, assignment and communication of responsibilities and authorities for performing the process and the commitment of the persons involved. The organization needs to ensure that all stakeholders in the process (e.g. process owner, process implementers, etc) are informed of these activities.

The organization must identify the resources and information needed to implement the process (in accordance with the identified process performance objectives). They must also ensure the resources and information are made available, allocated and used. The organization must be prepared to adjust the resources and information as part of the management of the process performance in response to deviations from the planned performance.

In the (normal) situation that the process involves different parties or teams within an organization or in external organizations, it is important that the interfaces are clearly defined and managed between the involved parties to ensure effective communication and clear assignment of responsibility. There are typically several types of stakeholders to consider – the process owner (s), the process implementer (s), the process resource provider(s) who provides the necessary resources and information, the clients and suppliers who are involved upstream and downstream of the process, and potentially others. Organizations often perform this activity in the initial planning but often fail to manage the effect of changes in the interfaces due to changes in process performance. Therefore it is vital that the interfaces between these parties be planned, monitored and adjusted as appropriate and that these be communicated in a clear and timely manner.

PA 2.2 Work product management attribute

Work products are artefacts associated with the execution of a process and hence the nature of the work product will vary depending on the purpose of the process and may be a part of the deliverable product or not (e.g. quality records).

The organization must manage the work products of the process (produced by the Capability Level 1 process) to provide reasonable assurance that they are appropriately identified, documented, and controlled.

This starts by identifying the requirements for the process work products to provide a basis for their production (as well as verification). These requirements can have a significant influence on the performance requirements for the process (e.g. the type and level of detail required to specify a web page is very different to specifying a Customer Relationship Management system). This simple example illustrates how the two process attributes at Capability Level 2 are interdependent.

The organization needs to cover both functional and non-functional requirements involved in managing work products. Functional requirements pertain to attributes of the work product (performance, size, etc.). Non-functional requirements pertain to agreements or constraints (e.g. delivery dates, packaging, etc.), which are not directly related to work product but to their production. Some requirements may be a combination of both functional and non-functional requirements. The requirements may be incorporated within a work product (for example in a template) or may be separate, for example in a checklist.

The organization must also define requirements for the documentation and control of the process work products, which are distinct from the requirements for the work products themselves. Depending upon the complexity of the work products, the complexity of the inter-relationships to other work products, the organizational unit specific or customer mandated needs; the organization must apply various degrees of change control or even configuration management.

The organization applies the defined requirements for the identification, documentation and control of the process work products.

To provide adequate product assurance, the organization must review the process work products resulting from implementation of the process, in accordance with the planned arrangements and adjusted as necessary to meet (changing) requirements. The extent and nature of the review will depend upon many factors (e.g. work product criticality and complexity, balancing resource use in review against production, balancing effort versus work product quality improvement, etc.) all of which should be considered as part of the planning for work product management.

State of Practice Experience

In the author's experience, organizations more often meet the performance management attribute than the work product management attribute. In cases of poor work product management the work product requirements are often well defined but problems occur due to the lack of update, change control and review. Organizations with strong review and configuration management processes often successfully fulfil the requirements of PA2.2.

Another interesting note is that many ISO 9000 organizations that met many of the Capability Level 3 requirements did not achieved a corresponding level of performance management of processes at level 2. This is sometimes due to inadequate education of those persons responsible for implementing and managing the processes (a lack of consistency), and is more evident in organizations without strong project management and review processes.

In addition, the organization involved in the performance management of processes varies according to the types of processes being performed. For example, engineering and support processes are often performed and managed within a project team, while organizational and the more general management processes (e.g. Human Resource Management, Quality Management, Improvement) may be performed and managed across an entire corporation or public body. When larger numbers of personnel are involved in a process, the way the process is defined, performed and managed often needs to be different to the way it is defined, performed and managed in a small group or project team.

In general the organizational and general management processes are less well performed than engineering and support processes, both due to the number of people involved across an organization and sometimes due to a clear lack of line management involvement and responsibility and unclear lines of authority.

Assessors need to consider consistency (or inconsistency) of process performance as the organizational scope changes (small to large number of people, number of organizational units involved, etc.).

Finally, on the positive side, PA2.1 is often fulfilled for processes in a project via regular progress reporting by the Project/Team Manager to their management.

Guidance to users of earlier versions of ISO/IEC 15504 Technical Reports

The description has been redrafted to better align with ISO 9000. The changes include

- adjusting the performance of the process to meet plans,
- resources and information are identified, made available and used (now part of PA 2.1 for specific project/instances whereas this was formerly specified explicitly only in PA 3.2),
- interfaces between involved parties are managed, and
- inclusion of qualitative and quantitative data for management purposes at level 2.

It was previously unclear whether quantitative data was a requirement for Capability Level 2 or was an indicator for Capability Level 4. The new version makes it clear that this quantitative data should be used, if appropriate, at Capability Level 2 and could include basic metrics covering cost and schedule tracking of the use of resources in a particular project/instance. Qualitative measures will still dominate and may be considered sufficient at this level (e.g. lessons learnt reports, ease of use feedback reports) for many processes. The use of quantitative data is considered project or instance specific at Capability Level 2, not institutionalised at a broader organization level, which is the intent at Capability Level 4.

There is now explicit allocation of resources , and the definition of responsibilities and authority for personnel allocated to projects/instances. This was previously only explicitly stated in PA 3.2 where it applies for defining and deploying processes throughout the organization.

Level 3: Established process

Level 3: Established process

The previously described *Managed process* is now implemented using a defined process that is capable of achieving its process outcomes.

The following attributes of the process, together with the previously defined attributes, demonstrate the achievement of this level.

5.4.1 PA 3.1 Process definition attribute

The process definition attribute is a measure of the extent to which a standard process is maintained to support the deployment of the defined process. As a result of full achievement of this attribute:

a) a standard process, including appropriate tailoring guidelines, is defined that describes the fundamental elements that must be incorporated into a defined process;

b) the sequence and interaction of the standard process with other processes is determined;

c) required competencies and roles for performing a process are identified as part of the standard process;

d) required infrastructure and work environment for performing a process are identified as part of the standard process;

e) suitable methods for monitoring the effectiveness and suitability of the process are determined.

NOTE A standard process may be used as-is when deploying a defined process, in which case tailoring guidelines would not be necessary.

5.4.2 PA 3.2 Process deployment attribute

The process deployment attribute is a measure of the extent to which the standard process is effectively deployed as a defined process to achieve its process outcomes. As a result of full achievement of this attribute:

a) a defined process is deployed based upon an appropriately selected and/or tailored standard process;

b) required roles, responsibilities and authorities for performing the defined process are assigned and communicated;

c) personnel performing the defined process are competent on the basis of appropriate education, training, and experience;

d) required resources and information necessary for performing the defined process are made available, allocated and used;

e) required infrastructure and work environment for performing the defined process are made available, managed and maintained;

f) appropriate data are collected and analysed as a basis for understanding the behaviour of, and to demonstrate the suitability and effectiveness of the process, and to evaluate where continuous improvement of the process can be made.

NOTE Competency results from a combination of knowledge, skills and personal attributes that are gained through education, training and experience.

[ISO/IEC 15504-2 5.4]

Clarification

At Capability Level 3, the organization has established and maintains a standard process. The organization then establishes a defined process, which is an instantiation of the standard process, to achieve its process outcomes for a particular context (for example in a project or in a department). The defined process may be a tailoring from the standard process to suit the requirements of the particular context (e.g. a simple or a complex project) or may use the standard process when this is suitable to use without change.

The definition and use of standard and defined processes is the primary distinction from the Managed Level (level 2).

The organization identifies resources – both human and infrastructure – needed for performance of the standard process. The resources are incorporated into the defined process. The organization collects appropriate data (qualitative and/or quantitative) on the performance of the defined process to identify opportunities for understanding and improving both the defined process and consequently the standard process.

The standard process which is tailored and effectively deployed along with the infrastructure needed to provide the basis for a closed loop feedback cycle for process improvement is the foundation required for effective application of process measurement at Capability Level 4.

PA 3.1 Process definition attribute

The process definition attribute is concerned with establishment of a standard process, its use as the basis for performance of the defined process and the collection and evaluation of process performance data as the basis for understanding and improvement of the standard process.

There is an increase in the scope of this process attribute, in that the definition of the process must also consider the sequence and interaction with other processes. In addition, the roles and competencies must be identified for the standard process. There must also be a monitoring of process effectiveness and suitability.

The organization must document the standard process and associated tailoring guidelines and make these available and known to those implementing the process. The organization's standard processes may be described at a general level that may not be directly usable to perform a process. This is not mandatory; it is possible for the standard process to be directly useable as a defined process, but this normally requires a high level of organizational discipline to ensure it is accurately implemented and no informal, unauthorized changes are implemented.

The tailoring guidelines need to provide clear direction regarding appropriate adaptation of the standard process to the range of applications for which the standard process is intended to apply. This should take into account the objectives, constraints and conditions that constitute the environment in which the process will be deployed (e.g. work scope, number of persons in the team). They should describe what can and cannot be modified and identify process components that are candidates for modification.

The resultant defined process is a maintained process description and provides a basis for planning, performing, and improving the organization unit's tasks and activities.

In addition, the defined process should contribute work products, measures, and other process improvement information to the organization's process assets (i.e. provide feedback to the standard processes). It can only do this if a mechanism is established in which the persons implementing the process act in accordance in the defined process (i.e. they implement the process with fidelity). As process usage data (qualitative and/or quantitative) are collected, the organization is able to create a basis for evaluating the behaviour of the standard process. This repository of knowledge provides the basis for understanding and (continual) improvement of the standard process.

PA 3.2 Process deployment attribute

The process deployment attribute is concerned with the effective deployment of a defined (and tailored as appropriate) process within an organizational unit.

When an organization achieves the process deployment attribute, the defined process as implemented retains fidelity to the standard process, so that the data on the deployment provides applicable feedback to the standard process.

The organization must ensure that enabling conditions for successful deployment (implementation) of the defined process are present. Enabling conditions include:

- Ensuring the people who implement the process have the required specific competencies. These competencies may either be based on appropriate education, training, skills and/or experience, and if necessary personnel are provided with appropriate training (and assistance) to obtain such competency.

- Understanding the process infrastructure required for performing the defined process.
- Successful allocation and deployment of the required human resources and process infrastructures.
- The people understand the roles, responsibilities and competencies for performing the defined process, they are given the appropriate authority to perform, manage and take on the responsibilities of the required tasks and outcomes, and management obtains their commitment to their assignment.

The organization must also effectively deploy the resources and information required to implement the defined process and ensure that these are used.

The organization must deploy an appropriate process infrastructure, which encompasses tools, methods and special facilities that are required for performance of the defined process.

The organization must determine, collect and evaluate appropriate data relating to implementation of the defined process in order to provide a basis for understanding the behaviour of the defined process as well as demonstrating the suitability and effectiveness of the defined process. This, in turn, contributes to the ongoing improvement of the standard process elements upon which the defined process is based.

State of Practice Experience

Even though the process attribute PA3.2 has changed in the new standard, experience is still relevant today. In the author's experience, ISO 9000 certified organizations often meet the process definition attribute but fail to meet the process deployment attribute. This was evidenced in several ways: often there were inadequate resources, they were not always sufficiently competent (due to lack of appropriate training), infrastructure and information was missing or never properly communicated, the defined process was not checked to ensure it was properly deployed and in use, and due to changing organizational or customer requirements, the deployed (implemented) processes were no longer appropriate.

Tailoring of processes was also poorly performed and feedback from the (tailored) defined process to the standard process was not evident.

On the positive side, organizations with strong quality assurance involvement in project teams and quality management involvement in project reviews collect feedback from the defined processes and use this to improve the standard processes and thereby achieve the requirements of PA3.1.

Guidance to users of earlier versions of ISO/IEC 15504 Technical Reports

The new version changes the attributes into process definition and process deployment as two integral aspects of the same overall management capability for processes. In simple terms, the former defines the (standard) process and the latter implements this as a defined process. This is more consistent than the previous version and clarifies the conflict that existed with the need for resources to be available and managed at level 2. Now these resources shall be defined for the standard and implemented processes and deployed in the controlled manner as defined.

The earlier version of ISO/IEC 15504 PA3.2 described a process resource attribute. This has been realigned in the new version of ISO/IEC 15504 into level 2 and level 3 as outcomes of their process attributes.

Other important additions include:

• the sequence and interaction of the standard process with other processes is determined,

• the required competencies, roles, responsibilities and authorities for performing a defined process are now also identified explicitly as part of the standard process,

• there must also be a monitoring of process effectiveness and suitability, and

• the required infrastructure and work environment for performing a defined process are explicitly identified as part of the standard process.

These additions strengthen the importance of the organization having standard processes (this also strengthens the need to handle process establishment and maintenance as an organizational process).

Part of the weakness of the precise definition of the prior process outcomes has been addressed in the new version, which will hopefully enable organizations to better meet the Capability Level 3 outcomes.

The basis for competence is explained further, the inclusion of information as a particular form of resource is described, and the definition of appropriate data to demonstrate suitability and effectiveness of the processes is emphasized.

The process deployment outcomes reflect the implementation of the process definition outcomes, emphasizing that defining these outcomes is not sufficient in itself, they must be put into practice (something that does not always occur as planned).

Level 4: Predictable process

Level 4: Predictable process

The previously described *Established process* now operates within defined limits to achieve its process outcomes.

The following attributes of the process, together with the previously defined attributes, demonstrate the achievement of this level.

PA 4.1 Process measurement attribute

The process measurement attribute is a measure of the extent to which measurement results are used to ensure that performance of the process supports the achievement of relevant process performance objectives in support of defined business goals. As a result of full achievement of this attribute:

a) process information needs in support of relevant defined business goals are established;

b) process measurement objectives are derived from process information needs;

c) quantitative objectives for process performance in support of relevant business goals are established;

d) measures and frequency of measurement are identified and defined in line with process measurement objectives and quantitative objectives for process performance;

e) results of measurement are collected, analysed and reported in order to monitor the extent to which the quantitative objectives for process performance are met;

f) measurement results are used to characterize process performance.

NOTE 1 Information needs typically reflect management, technical, project, process or product needs.

NOTE 2 Measures may be either process measures or product measures or both.

PA 4.2 Process control attribute

The process control attribute is a measure of the extent to which the process is quantitatively managed to produce a process that is stable, capable, and predictable within defined limits. As a result of full achievement of this attribute:

a) analysis and control techniques are determined and applied where applicable;

b) control limits of variation are established for normal process performance;

c) measurement data are analysed for special causes of variation;

d) corrective actions are taken to address special causes of variation;

e) control limits are re-established (as necessary) following corrective action.

[ISO/IEC 15504-2 5.5]

Clarification

At level 4, the organization consistently operates its processes within defined process control limits that aid the predictability of achievement of the process outcomes. The organization must use quantitative measurement (of both process and related product measures) to manage (control), monitor and set performance criteria for deployment (implementation) of defined processes. The quantitative management of the process supports the overall business goals.

Process measurement and control now occur at wider organizational levels, for example a division running multiple projects would collect and use measurements across several related projects or maybe even all its projects as a means to assess the capability of processes to support the overall business goals of that division.

When performance falls outside the defined limits, the reasons must be investigated and assigned to special causes[17]. The causes of variation in performance must be addressed and actions taken to adjust process performance back to the defined limits. Both the standard and the defined process may be affected by the actions.

PA 4.1 Process measurement attribute

The organization must implement an effective system for the collection of measures relevant to the performance of the process and the quality of the work products.

In order to do this, the organization must clearly identify and describe the overall business goals and then the corresponding organizational unit's specific goals and measures for product and process. This requires the organization to identify suitable quantitative process performance objectives and related quantitative product and process measures that measure the extent of achievement of the organization's business goals.

The organization applies and evaluates the quantitative measures to the performance of the process to determine the extent of achievement of the organization's business goals. Measures of process performance can include cost and schedule, as well as product quality and productivity.

The organization shall perform measurement of the implemented processes and relate this to the standard processes. This may include capability

---

[17] Special causes of variation refer to defects in a process, which are not inherent to the process but rather are incidental; these typically stem from implementation problems.

determination using process assessment, for example, by use of this standard) or by evaluation of the consistency of performance (in other words, the ability to control performance within defined limits).

The organization maintains records of measurements of processes, both for historical comparison and for ongoing management and improvement of the process.

The organization may evaluate trends in performance as a form of analysis with the aim to confirm the adequacy and suitability of measures, and to help identify common and special causes of process variation.

PA 4.2 Process control attribute

The organization must choose and identify analysis and control techniques that are relevant to the nature of the process as well as the overall context of the organizational unit being assessed. This must also include the requirements for the frequency of measurements.

The organization must define acceptable control limits for process performance (based on business goals and the related organizational unit goals). In general, processes vary in performance and this variation is called common cause variation[18]. This variation may be caused by people or defined process variations (different process implementations to suit different circumstances). The control limits can be defined either based on experience (these may initially be estimated or based on past data), or defined in terms of establishing targets for performance. The control limits must recognise that common cause variations occur and be realistic in what is the amount of controllable variation. It does not make sense to set control limits that are impossible to meet due to process variation.

The organization applies the analysis techniques that have been identified for identifying the root causes of variation in process performance (and whether they are common or special causes).

Where the root causes of variations create process instability beyond the acceptable control limits, the organization must implement effective corrective action(s) designed to address the identified root causes of variation. The organization must use measurement and analysis to justify decisions taken, based on their impact on delivery of benefit to the business.

---

[18] Common causes of variation refer to normal ongoing process variations that are inherent to the process as defined.

The organization must re-establish process control limits after taking corrective action. The control limits may be different to those before the corrective action was taken.

Note: not all processes are equally suited to statistical control, and alternative techniques may need to be selected that demonstrate a qualitative understanding of the process.

State of Practice Experience

In the author's experience, organizations that were fully achieving Capability Level 3 and attempting to fully achieve Capability Level 4 were good on establishing process measurement programs, but did not as often control the processes quantitatively.

The main shortcomings involved relating the overall business goals to the organizational unit goals, setting acceptable process control limits, and using control charts or other techniques to systematically analyse process capability in a way that lead to detection of unacceptable process variation and resultant corrective actions.

In general, organizations with strong project management methodologies complemented by formal cost and schedule control systems were able to fully achieve the outcomes of the process attributes PA4.1 and PA4.2 for their project based processes and the related metrics of achievement of their business goals.

Guidance to users of earlier versions of ISO/IEC 15504 Technical Reports

The new version of the process outcomes of the process attributes clarifies the intent of the earlier versions of the standard. In particular, under process measurement, it first requires the organization to establish quantitative objectives for process performance in support of the defined business goals. These objectives then lead to quantitative objectives and identification of measurements. PA 4.1 also explicitly requires establishment of measurement infrastructure and assignment of responsibilities for this outcome. Process capability measurement is assumed to be one applicable form of process measurement.

For process control, the new version of the standard explicitly requires control limits of variation, control parameters for process performance, and frequency of measurements to be defined. In addition, it explicitly requires corrective actions to be taken in response to variations caused by special causes.

Level 5: Optimising process

Level 5: Optimising process
The previously described *Predictable process* is continuously improved to meet relevant current and projected business goals.
The following attributes of the process, together with the previously defined attributes, demonstrate the achievement of this level.
PA 5.1 Process innovation attribute
The process innovation attribute is a measure of the extent to which changes to the process are identified from analysis of common causes of variation in performance, and from investigations of innovative approaches to the definition and deployment of the process. As a result of full achievement of this attribute:
a) process improvement objectives for the process are defined that support the relevant business goals;
b) appropriate data are analysed to identify common causes of variations in process performance;
c) appropriate data are analysed to identify opportunities for best practice and innovation;
d) improvement opportunities derived from new technologies and process concepts are identified;
e) an implementation strategy is established to achieve the process improvement objectives.
PA 5.2 Process optimization attribute
The process optimization attribute is a measure of the extent to which changes to the definition, management and performance of the process result in effective impact that achieves the relevant process improvement objectives. As a result of full achievement of this attribute:
a) impact of all proposed changes is assessed against the objectives of the defined process and standard process;
b) implementation of all agreed changes is managed to ensure that any disruption to the process performance is understood and acted upon;
c) effectiveness of process change on the basis of actual performance is evaluated against the defined product requirements and process objectives to determine whether results are due to common or special causes.
[ISO/IEC 15504-2 5.6]

Clarification

When the organization achieves the optimising process Capability Level, it continually changes and adapts its processes in an orderly and intentional manner to effectively respond to changing business goals.

This level of process capability fundamentally depends on the quantitative understanding of process behaviour derived through a predictable process capability.

An organization operating an optimising process (Capability Level 5) has the following critical behaviours. First, the organization has a proactive focus on continuous (continual) improvement in the way that it fulfils current and projected business goals. The improvement efforts are both intentional and planned in order to improve the effectiveness and efficiency of the existing processes (process optimisation of the predictable, defined processes and standard processes).

Second, the organization has an orderly and planned approach to identifying appropriate changes to the existing process and for new processes (process innovation). The organization introduces them in a manner that minimizes undesired disruption to the operation of the process itself and/or the sequence and interaction with other processes.

Third, the organization evaluates the effectiveness of the changes against actual results achieved, and makes further adjustments as necessary to achieve desired product and process goals.

The organization establishes quantitative objectives (based on the relevant business goals of the organization) for improvement of process performance. It collects and analyses data to identify common causes of variation in performance and to identify opportunities for best practice and innovation. The organization pilot innovative ideas and technologies in processes in order to meet defined business and improvement goals or objectives.

PA 5.1 Process innovation attribute

The process innovation attribute is concerned with the existence of the organization's proactive focus on improvement in the fulfilment of both current and projected (relevant) business goals of the organization. These goals must lead to explicit defined process improvement goals.

The relevant business goals and process improvement goals are the drivers for all of the level 5 outcomes.

The organization must strive to understanding the source of existing process problems and variations (common cause variation), as well as any potential process problems induced by process improvement goals, and use this understanding as a source of proposed process changes. The organization assesses the proposed process changes to existing processes with respect to the current and projected (relevant) business goals in order to select the most appropriate changes to implement.

The organization needs to have a top-level strategy to ensure successful achievement of process innovation due to the inherent complexity of organizational deployment, the long-term nature of improvement and the situation that results often rely upon aggregate changes.

PA 5.2 Process optimisation attribute

The process optimisation attribute is concerned with an orderly and proactive approach to identifying appropriate changes to improve an existing process or set of related processes.

The organization should consider how best to optimise its use of resources in order to achieve the best improvements possible with available resources. It should estimate the impact of proposed changes based in part on the quantitative understanding of the predictable process (from Capability Level 4).

The organization must ensure that introduction of changes minimizes undesired disruption to the operation of the process. This involves the timing and sequencing of agreed changes, the organizational unit activity criticality and status, process change effectiveness evaluation and new business generation.

The organization must evaluate the effectiveness of changes against actual results and make adjustments as necessary to achieve the relevant process improvement objectives.

Finally, the organization must use the evaluation of the effectiveness of changes as the basis for continuous learning and ongoing process knowledge acquisition and management.

State of Practice Experience

In the author's experience, very few organizations have implemented effective Capability Level 5 processes. To implement processes at Capability Level 5 requires a high degree of ongoing organizational commitment (over a long period of time) to provide resources and promote process innovation and optimisation. In few cases has the author seen truly effective level 5 processes sustained over long periods, even in organizations claiming to be at level 5. On the other hand, those organizations that have achieved level 5 in at least some of their processes are often better able to adapt to changes in their competitive environment.

In situations that organizations have designed processes to be at Capability Level 5, the main problems have been demonstrating effective process improvements that continually improve the achievement of business goals. Process innovation and optimisation needs to result in quantitative cost, time-to-market, product quality and/or customer satisfaction improvements. When the demonstration of the improvements does not positively impact on the bottom line of the business, then top level organization management support can be easily lost, especially from managers not totally committed to quality improvement and driven by short term business profitability objectives.

All change requires additional effort and commitment. Therefore, another critical challenge is to minimize the undesired interruption caused by change, otherwise the persons (and especially managers) involved quickly lose a desire to make further changes.

Guidance to users of earlier versions of ISO/IEC 15504 Technical Reports

The process attributes have been completely renamed and the emphasis changed to reflect best practice management thinking and approaches. Overall, the goals are the same, but the outcomes are expanded.

The process innovation attribute was formerly called the process change attribute. The main changes here are to re-emphasize that innovations are related to process improvement objectives (which support the business goals); the data from the predictable processes is analysed (including the process variation data) in order to identify best practice within the organization as well as opportunities for process innovation; the impact of technology is assessed for potential process innovations; and an implementation strategy is established.

The process optimisation attribute was formerly called the continuous improvement attribute (the former PA 5.1 is substantially now PA 5.2). The main changes here are:

- to assess improvements against business goals and not just the objectives of the defined and standard processes;
- the implementation is managed to minimize disruption; and
- evaluation of process change assesses whether results are due to common or special causes.

4.3   Rating process attributes

In addition to the ordinal scale for measuring process capability, there is a scale for rating the extent of achievement of the process attributes.

Process attribute rating scale

The extent of achievement of a process attribute is measured using an ordinal scale of measurement as defined below.

Process attribute rating values

The ordinal rating scale defined below shall be used to express the levels of achievement of the process attributes.

N Not achieved.

There is little or no evidence of achievement of the defined attribute in the assessed process.

P Partially achieved.

There is some evidence of an approach to, and some achievement of, the defined attribute in the assessed process. Some aspects of achievement of the attribute may be unpredictable.

L Largely achieved.

There is evidence of a systematic approach to, and significant achievement of, the defined attribute in the assessed process. Some weakness related to this attribute may exist in the assessed process.

F Fully achieved.

There is evidence of a complete and systematic approach to, and full achievement of, the defined attribute in the assessed process. No significant weaknesses related to this attribute exist in the assessed process.

The ordinal points defined above shall be understood in terms of a percentage scale representing extent of
achievement.

The corresponding values shall be:

N	Not achieved	0 to 15% achievement
P	Partially achieved	16% to 50% achievement
L	Largely achieved	51% to 85% achievement
F	Fully achieved	86% to 100% achievement

[ISO/IEC 15504-2 5.7.1-5.7.2]

The assessors must use some expert judgment in rating the extent of process capability and the extent of fulfilment of a process attribute. As such, the above numerical values of the rating levels (anchor points) are intended to provide guidance to assessors rather than be explicitly recorded. The nonlinear anchor point placement in the scale is intended to guide the assessor's discriminatory judgment when rating the process attribute.

The assessors also need to consider the relative importance of each of the outcomes of the process attributes for the organizational unit assessed. Not all outcomes may be of equal importance and this may also vary according to the environment in which the process is used. This expert judgment should be considered against the ordinal anchor points in rating the process attribute fulfilment. Similarly, the assessed organization should not only consider the rating, but also what outcomes were/were not achieved in order to assess their importance to the business goals.

Process attribute ratings

Each process attribute shall be rated using the ordinal rating scale defined above. A process shall be assessed up to and including the highest capability level defined in the assessment scope.

NOTE The set of process attribute ratings for a process forms the process profile for that process. The output of an assessment includes the set of process profiles for all assessed processes.

Referencing of process attribute ratings

Each process attribute rating shall be given an identifier that records the process name and the process attribute assessed.

NOTE The ratings may be represented in any format, such as a matrix or as part of a database, provided that the representation allows the identification of individual ratings according to this referencing scheme.

[ISO/IEC 15504-2 5.7.3-5.7.4]

The assessors must follow a formalized Process Assessment Model in order to assure that the process attribute ratings are verifiable.

The assessors need to present the process attribute ratings for each assessed instance of a process. The resultant set of ratings of the process attributes is termed the process profile for that process (i.e. the measurement of capability). The process profile may contain up to nine ratings (one for each process attribute) depending on the Capability Level assessed.

Any Process Assessment Model used for assessment should provide a mechanism for expressing the evaluation of process capability as a series of process profiles.

4.4 Process Capability Level model

Achievement of Capability Levels

The assessor shall derive the Capability Level of a process from the process attribute ratings according to the Process Capability Level model defined in the Capability level-rating table.

In brief, this means that for a process to be rated at a particular Capability Level, it must be largely or fully achieved at that level and all lower Capability Levels must be fully achieved.

Capability level ratings

Scale	Process Attributes	Rating
Level 1	Process Performance	Largely or fully
Level 2	Process Performance	Fully
	Performance Management	Largely or fully
	Work Product Management	Largely or fully
Level 3	Process Performance	Fully
	Performance Management	Fully
	Work Product Management	Fully
	Process Definition	Largely or fully
	Process Deployment	Largely or fully
Level 4	Process Performance	Fully
	Performance Management	Fully
	Work Product Management	Fully
	Process Definition	Fully
	Process Deployment	Fully
	Process Measurement	Largely or fully
	Process Control	Largely or fully
Level 5	Process Performance	Fully
	Performance Management	Fully
	Work Product Management	Fully
	Process Definition	Fully
	Process Deployment	Fully
	Process Measurement	Fully
	Process Control	Fully
	Process Innovation	Largely or fully
	Process Optimisation	Largely or fully

As can be seen in the table, each lower Capability Level must be fully achieved and can no longer be only largely achieved.

For example, a process rated at Capability Level 3 can have the process attributes: Process Definition and Process Deployment only rated as largely achieved (51 to 85% achievement), but if this process is to be rated at Capability Level 4, then the process attributes: Process Definition and Process Deployment must be rated as fully achieved (86 to 100% achievement).

The reason to require all lower levels to be fully achieved in order to rate at a higher level is that the design of the Capability Levels is accumulative and sustaining – the lower level practices need to be present to ensure that the organization consistently implements, manages and applies the practices, so that the cumulative practice achieves better results for the organization.

Theoretically, at the higher Capability Levels, it is still possible to have several small weaknesses at each of the lower Capability Levels since a rating of 'fully achieved' means between 86 to 100% achievement. In many cases, too many small weaknesses at lower Capability Levels tend to lead to (or reflect) more important management weaknesses at the higher Capability Levels due to the cumulative nature of the practices. However, it is possible to have some minor weaknesses at the lower Capability Levels and assessors need to take note of these in an assessment and use their expert judgment when rating the processes at the higher Capability Levels.

The following diagram illustrates graphically an example assessment result.

The assessed processes are listed on the left hand side, for example, CUS.3 Requirements Elicitation. The 9 process attributes are shown at the top in order from PA1.1 to PA 5.2[19].

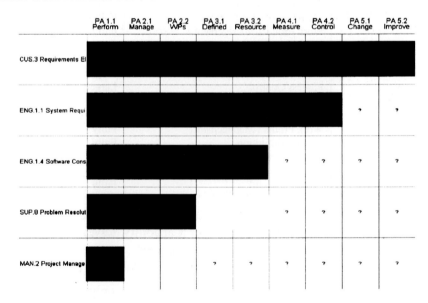

Example Capability Level ratings – SPICE 1-2-1 representation.

---

[19] This representation is used by SPICE 1-2-1. Other representations are possible.

Using the interpretation from the Capability Level ratings table, the example Capability Level ratings as shown using SPICE 1-2-1 are:

- CUS.3 is Capability Level 5 (all process attributes from PA1.1 to PA5.2 are fully achieved).
- ENG 1.1 is Capability Level 4 (PA1.1 to PA3.2 are fully achieved and PA 4.1 and PA 4.2 are largely achieved).
- ENG 1.4 is Capability Level 3 (PA 1.1 to PA3.2 are fully achieved).
- SUP.8 is Capability Level 2 (PA1.1 and PA2.1 are fully achieved and PA2.2 is largely achieved).
- MAN.2 is Capability Level 1 (only PA1.1 is fully achieved, no other PA is largely achieved).

4.5   Compatibility with ISO 9000:2000 Requirements

The ISO 9000 family of standards [3] has advocated quality management systems in organizations in a wide range of industries with a focus on customer satisfaction. The first release of ISO 9001 was in 1994 [4].

ISO 9001 specifies requirements on quality management systems including requirements for quality assurance and quality control. The way these have been specified has evolved over time. The ISO 9000 standards have been influenced by the development of process-related standards such as ISO/IEC TR 15504:1998, ISO/IEC 12207:1995 and ISO/IEC 15288:2001. As a result, ISO 9001:2000 has adopted a process oriented approach to product development and service delivery. This is expressed in ISO 9001:2000 as: *"promotes the adoption of a process approach when developing, implementing and improving the effectiveness of a quality management system, to enhance customer satisfaction by meeting customer requirements."*

ISO 9001 is a small document, using an abstract model focused on certification. In ISO 9001, the process-oriented approach requires only 6 documented procedures and it focuses on quality records.

ISO 9004 provides further guidance for an organization's quality management system objectives: *"particularly for the continual improvement of an organization's overall performance and efficiency, as well as its effectiveness"*.

It is an informative part of the standard, and provides guidelines for improvement and self assessment that allow:

- application to the entire quality management system, or to a part, or to any process,
- application to the entire or part of the organization,
- identification and prioritisation of opportunities for improvement, and
- facilitation towards **maturing**[20] of the quality management system towards world-class performance.

In Annex A of ISO 9004, it specifies a five performance maturity levels, different to those in ISO/IEC 15504. It is clear that ISO 9000 in the ISO 9004 guidelines recognizes and promotes organizations to embrace an approach that recognizes a performance dimension to its process performance, even if the described levels are formulated differently to ISO/IEC 15504. These were taken into account in the latest revision of ISO/IEC 15504 although they are still not compatible[21].

---

[20] Author's emphasis.

[21] There are ongoing discussions on what are the best capability or maturity level definitions. The definition – 'world's best practice' would probably be difficult to rate as it is almost impossible to define what is the best practice in the world, it varies over time and industry/regional/national cultures affect the way practices are implemented.

ISO 9004:2000 Performance Maturity Levels and ISO/IEC 15504-2:2003

Maturity Level	Performance Level	Capability Level	Performance Level
		0	Incomplete The process is not implemented, or fails to achieve its process purpose.
1	No formal approach No systematic approach evident, no results, poor results or unpredictable results.	1	Performed The implemented process achieves its process purpose.
2	Reactive Approach Problem or corrective-based systematic approach; minimum data on improvement results available.	2	Managed The process is now implemented in a managed fashion.
3	Stable Formal System Approach Systematic process-based approach, early stage of systematic improvements; data available on conformance to objectives and existence of improvement trends.	3	Established The process is now implemented using a defined process that is based upon a standard process and that is capable of achieving its process outcomes.
4	Continual improvement emphasized Improvement process in use; good results and sustained improvement trends.	4	Predictable The process now operates within defined limits to achieve its process outcomes.
5	Best-in-class performance Strongly integrated improvement process; Best-in-class benchmarked results demonstrated.	5	Optimising The process is continuously improved to meet relevant current and projected business goals.

As can be seen in the comparison table [5], the levels are not directly compatible but in general are focused on demonstrable improved levels of capability from levels 1 to 5. The level scale in ISO/IEC 15504 is of finer granularity than ISO 9004 (6 levels instead of 5) and does not prescribe the ultimate 'Best-in-class' performance maturity level of ISO 9004. In very general terms:

- Level 1 in ISO 9004 is broadly equivalent to level 0 in ISO/IEC 15504,
- Level 2 in ISO 9004 is between level 1 and 2 in ISO/IEC 15504, and
- Level 3 is broadly equivalent in both.

- Level 4 and level 5 allocate the use of statistical process control, metrics and improvement practices differently and are hard to correlate to each other.

Note that the normative part of the standard, ISO 9001:2000 primarily focuses on performance maturity levels 1 to 3 – we will see how that translates later in this section.

Despite the differences in the informative ISO 9004, the widespread acceptance of ISO 9001 and its process orientation provides a natural candidate architecture for defining a Process Assessment Model.

A project sponsored by the European Space Agency has developed a quality management assessment model based on the quality management system requirements of ISO 9001, the guidelines in ISO 9004 in part, and the capability requirements of ISO/IEC 15504.

The process dimension of this overall Quality Management standard includes quality assurance and quality control requirements with an objective to provide adequate confidence to the customer that the end product or service satisfies the requirements. This can comprise activities covering adequate specification and fulfilment of product requirements including controls for the procurement, design, fabrication, integration, test and maintenance of products, maintenance of quality records and nonconformances. It also comprises activities related to interfaces with customers as well as suppliers.

In order to establish an assessment model, all the requirements from ISO 9001 were translated into practices. Practices are categorized as *generic practices*, i.e. those applicable to all processes, or *process-specific practices*, those practices only applicable to a particular process.

For creation of a conformant process reference and assessment model, the generic and process-specific practices were systematically associated with the appropriate process attributes. Process specific practices associated with performance of a process are indicators of process performance for the process dimension and can be considered as (mandatory) base practices. In addition, indicators include work products and work product characteristics. These indicators were assigned as the inputs or outputs of the processes.

It is possible to do the translation from requirements to processes and process practices. One possible translation into a Process Reference Model includes the following process categories and processes.

Customer Focus
Requirements Determination
Customer Communication
Customer Property
Product Acceptance & Delivery

Organisation
Management Direction
Quality Establishment
Management Review
Document Control
Record Control

Core Processes
Design & Development Management
Design & Development Control
Technical Review
Production and Service Provision
Control
Product Preservation
Design & Development
ChangeManagement
Product Identification

Resource and Facilities Management
Infrastructure Management
Work Environment Management
Human Resource Management
Control & Monitoring of Measuring
Devices

**Measurement, Analysis &
Improvement**
Internal Audit
Process Monitoring and Measurement
Product Monitoring and Measurement
Non-conforming Product Control
Corrective Action
Preventive Action
Continual Improvement

Purchasing Assurance
Supplier Selection
Purchasing Requirements
Supplier Monitoring
Incoming Inspection

Compatible ISO 9000 process categories and processes.

The resultant Process Reference Model is compatible with the requirements of ISO/IEC 15504.

The generic practices derived from ISO 9001:2000 associated with different process attributes support the revised definition of the ISO/IEC 15504 process attributes. Since ISO 9001:2000 does not specify the performance maturity (capability) dimension (rather only the informative ISO 9004 does); it is valid to consider the use of the ISO/IEC 15504 capability dimension. Therefore using the ISO/IEC 15504 capability dimension, the normative parts of the standard in ISO 9001:2000 can be represented as follows.

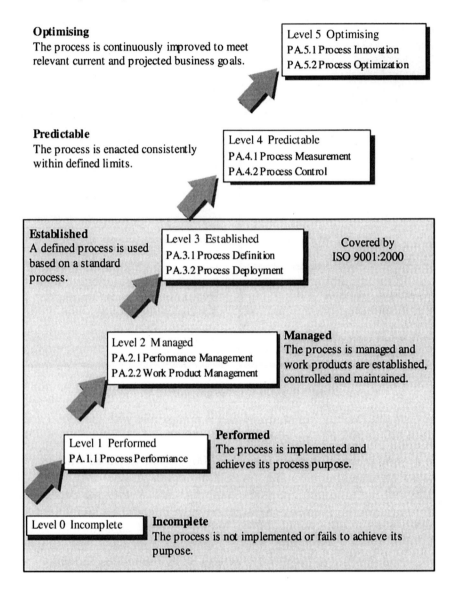

Optimising
The process is continuously improved to meet relevant current and projected business goals.

Level 5 Optimising
PA.5.1 Process Innovation
PA.5.2 Process Optimization

Predictable
The process is enacted consistently within defined limits.

Level 4 Predictable
PA.4.1 Process Measurement
PA.4.2 Process Control

Established
A defined process is used based on a standard process.

Level 3 Established
PA.3.1 Process Definition
PA.3.2 Process Deployment

Covered by
ISO 9001:2000

Level 2 Managed
PA.2.1 Performance Management
PA.2.2 Work Product Management

Managed
The process is managed and work products are established, controlled and maintained.

Level 1 Performed
PA.1.1 Process Performance

Performed
The process is implemented and achieves its process purpose.

Level 0 Incomplete

Incomplete
The process is not implemented or fails to achieve its purpose.

ISO 9001:2000 and ISO/IEC 15504 Capability dimension.

Using the developed model and a compliant assessment method allows organizations to perform assessments to determine their capability of quality management processes (thereby verifying compliance with ISO 9001). The use of assessment, rather than audit, for determining the capability of the processes of a quality management system provides a valuable tool to all organizations interested in implementing, maintaining and improving a quality management system based on ISO 9001.

In summary, it is possible to have a conformant assessment model for ISO 9001:2000 using the ISO/IEC 15504 capability dimension (rather than the informative ISO 9004 performance maturity levels). Naturally, the proposed model and the capability dimension will be subject to further discussion and evolution within the communities of interest.

5 PROCESS REFERENCE MODELS

In this chapter, the process dimension of the standard is described. This is defined in Process Reference Models. In the latest version of the standard, these models are external to the standard so that an organization can choose the most appropriate model to use.

The chapter provides an overview of the generic requirements and then illustrates this with several examples of conformant models in international standards, and some industry specific models.

It is important to note that Process Reference Models are primarily designed to guide users in what processes to implement (rather than assess). The models often provide a framework with implied process relationships. For example, software design may be related to software requirements analysis and software coding. The models often group processes into categories, such as development, management, and support categories as a guide to users about potentially related processes (process chains). These can be useful from the viewpoint of the person/team interested in implementing these processes.

Some models describe the framework in terms of activities and tasks, rather than process purpose and outcomes. Therefore, not all Process Reference Models may be compatible with the needs of process assessment, and may require translation/mapping. For ISO standards, this is part of the work of the ISO harmonisation project. For all other models, the community of interest needs to perform the required translation/mapping.

It is also important to know that the Process Reference Models in the international standards do not attempt to describe 'how' to implement the processes. (Note: Process Lifecycle models in a later chapter cover this). By avoiding the 'how' aspects, Process Reference Models can be managed to both provide the user with both process implementation and the process assessment aspects.

This chapter will be of interest to anyone designing process implementations, particularly process owners, process experts, and developers of models.

5.1 Requirements on Process Reference Models

One of the two dimensions of the ISO/IEC 15504 process assessment standard is the process dimension. This is as equally important as the capability dimension. The process dimension must provide clear process descriptions.

In order to assess an organization's business processes, the processes must be defined in a way that can be assessed. To handle this aspect, the standard refers to the need for a Process Assessment Model, which shall be based upon a Process Reference Model.

A compliant or conformant Process Reference Model defines a set of processes, and consists of statements of process purpose and expected process outcomes. The process purpose describes the high-level objectives of the process, while the process outcomes are the expected results of a successful implementation of the process. Together they describe what to achieve, without prescribing how to achieve the objectives.

Since the standard was initially devised as a software process assessment standard, it has a strong relationship with software lifecycle process standards, particularly ISO/IEC 12207. In the earlier technical report form of the standard, the Process Reference Model and Process Assessment Models were directly based upon ISO/IEC 12207 and incorporated within the standard and its informative parts.

ISO/IEC TR 15504-2:1998 (part 2) described the Process Reference Model. ISO/IEC TR 15504-5 (part 5) described an exemplar (example) Process Assessment Model.

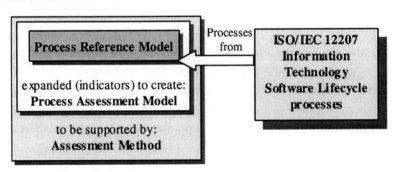

ISO/IEC TR 15504 (Technical Report) Process Reference and Assessment model.

In the latest version of ISO/IEC 15504-2, the standard no longer contains a Process Reference Model but states the requirements for defined Process Reference Models and Process Assessment Models, which are related to the measurement framework defined in the standard.

The Process Reference Model is now defined externally to ISO/IEC 15504 and must provide the basis for one or more Process Assessment Models.

The Process Assessment Model(s) are based on the process descriptions provided in Process Reference Model(s).

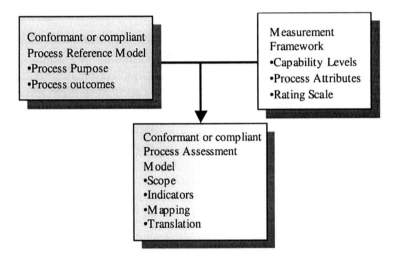

ISO/IEC 15504 Process Reference and Assessment Model relationships.

This second conformance requirement should enable organizations to compare assessment outputs from different Process Assessment Models that are based on the same Process Reference Model. In other words, an assessor should be able to obtain the same assessment result using different Process Assessment Models.

It should be noted that the expanded detail required for a conformant Process Assessment Model is substantial and not a trivial exercise. This probably means that very few different conformant models are likely to be available.

For a conformant Process Assessment Model, the term Translation, as used in the above figure, is concerned with assuring that assessment results can be converted into an ISO/IEC 15504 process profile in a repeatable and reliable manner. To assure this, the Process Reference Model(s) shall adhere to the following requirements:

a) The is a clear declaration of the domain of the Process Reference Model. For example, it may be for systems, or software or even more specific domains such as space software development.

b) The descriptions of the processes within the scope of the Process Reference Model meet the requirements of clause 6.2.4 of ISO/IEC 15504-2 (with regard to purpose and outcomes).

c) A description of the relationship between the Process Reference Model and its intended context of use (for example, the type and size of organization using it).

d) There is a description of the relationship between the processes defined within the Process Reference Model (for example, are processes grouped in a particular category such as customer-supplier processes?).

e) The Process Reference Model shall document the community of interest of the model, and also document the actions taken to achieve consensus within that community of interest.

   1) The relevant community of interest will be characterized and/or specified (for example, space industry, and automotive industry).

   2) The extent of achievement of consensus shall be documented; or

   3) If no actions are taken to achieve consensus, a statement to this effect shall be documented.

f) The processes defined within a Process Reference Model shall have unique process descriptions and identification.

For ISO/IEC 15504 purposes, any other elements in a Process Reference Model are to be considered informative, but for their users this information is still important in helping interpret the model and its application.

For ISO/IEC 15504 purposes, the fundamental elements of a Process Reference Model are the set of descriptions of the processes. These process descriptions shall meet the following requirements:

a) A process shall be described in terms of its purpose. The purpose states the high level overall objective of performing the process, and describes the likely outcomes of effective implementation.

b) In any description, the set of process outcomes shall be necessary and sufficient to demonstrate successful achievement of the process's purpose. The outcomes could be:
   1) Production of an artefact (e.g. system, software, document).

2) A significant change of state.

3) Meeting of specified constraints, e.g. requirements, goals etc.

c) The process descriptions shall describe the basic processes and not the process capabilities beyond level 1 for ISO/IEC 15504. In other words, they shall describe the process dimension and not the capability dimension.

What does all the above mean?

Basically, ISO/IEC 15504 now allows any type of Process Reference Model to be used, not just a software Process Reference Model.

Therefore, in the future it will be possible to use a model that describes the complete processes of an organization, not just the information technology or software processes (for example, a Process Reference Model covering ISO 9000 can be defined and used). This is an intentional result of the overall harmonization aspect of the standards setting approach for ISO/IEC 15504.

Furthermore, organizations and industry associations that perceive a strategic competitive advantage in their business processes, now have the option to create their own Process Reference Model or use an existing process model they already have. They can then create a Process Assessment Model based on this reference model.

However, the effort needed to ensure that the models are compliant or conformant with ISO/IEC 15504 should not be underestimated [22]. It is likely that only large organizations will undertake the step. The European space and automotive industries are taking this step through their industry associations.

Without the steps to ensure conformance, the organization can still assess its own processes using just the capability dimension and its own process descriptions, but this is not considered a conformant assessment.

Several Process Reference Model already generally meet the conformant Process Reference Model requirements. At this time most are conformant

---

[22] ISO defines compliance as: "adherence to those requirements contained in standards and technical reports which specify requirements to be fulfilled by other standards, technical reports or ISPs (e.g. reference models and methodologies)." ISO define conformance as "Conformity is fulfilment by a product, process or service of specified requirements."

with ISO/IEC TR 15504:1998 but many are being considered for revision or are being revised in line with ISO/IEC 15504 [23].

They include:

- ISO/IEC 12207 Amd.1: 2002 Information technology - Software Lifecycle processes.
- ISO/IEC 15288:2002 Systems Engineering - System Lifecycle processes.
- S9K.
- The Software Engineering Institute CMMI® and the FAA-iCMM®.
- Trillium© [24].
- Automotive SPICE ®.

The Software Engineering Institute CMMI®   and the FAA-iCMM® encapsulate the reference model within their overall model. The process parts of these documents in general meet the requirements for a Process Reference Model, but the overall model descriptions also include the capability dimension, which do not strictly comply with the ISO/IEC 15504 requirements. Rather than describe these models here, they are covered in chapter 6.

5.2   Information Technology – Software Lifecycle Processes

First, let us look at the standard that provided much of the process dimension and domain of application for ISO/IEC 15504. This is ISO/IEC 12207:1995 Information Technology - Software Lifecycle Processes, which established a common framework for software life cycle processes.

The evolution of the software lifecycle process standard and the process assessment standard illustrates many of the characteristics of standards setting including:

---

[23] At the time of writing of this book, ISO/IEC 15504 was being formally accepted as the standard and there is a transition period required to update these models. The author expects the transition period will vary depending upon the model from at least 1 year onwards.

[24] Trillium© is a proprietary process assessment standard used by Bell Canada to assess the product development and support capability of prospective and existing suppliers to Bell Canada for IT and telecommunications products. It will not be covered further in this book. See Coallier, F. (and others). The Trillium Model. Bell Canada for further details, and http//ricis.cl.uh.edu/trillium.

- Dependencies between standards.
- The way one standard moves ahead of the other ('leapfrogging').
- How ISO/IEC 15504, while starting as a software process assessment standard, has changed into a more generic process assessment that still integrates with ISO/IEC 12207.

The substantive part of the ISO/IEC 12207:1995 standard sets out for each process the purpose, activities and tasks required to implement the high-level life cycle processes to achieve desirable performance. The combined purpose, activities and tasks thereby specify the requirements for acquirers, suppliers, developers, maintainers and operators of software systems.

The standard groups the purposes and outcomes in a general process architecture within three life cycle process categories: Organizational, Primary and Supporting.

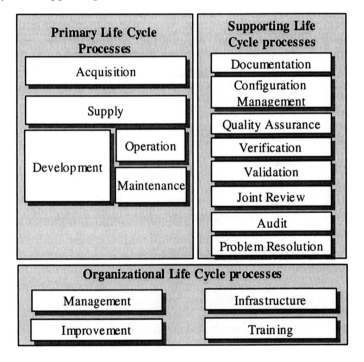

ISO/IEC 12207:1995 Process Reference Model.

Each process category contains a set or processes and sub-processes with purpose statement descriptions, comprising unique functional objectives when instantiated (defined and performed) in a particular environment. The purpose statement includes additional material identifying the outcomes of successful implementation.

The ISO/IEC 12207:1995 standard does not provide a complete description of a process life cycle, for example it does not describe phasing or dependencies between the processes. It must be elaborated to provide a complete process life cycle description. There is also a chapter on tailoring the standard to suit the organization.

When an organization applies the standard, it can choose to fully conform or partially conform. For full conformance, the organization declares the (standard's) processes for which full conformance is claimed and must be able to demonstrate that all the requirements have been met. For partial conformance, tailoring shall follow the guidelines in Annex A of the standard.

The ISO/IEC 15504 working group took the ISO/IEC 12207 reference model as a common basis for developing software process assessment. During this development, several issues were highlighted concerning processes and the granularity of the process definitions (degree of detail) in ISO/IEC 12207. During the SPICE trials, assessors found it difficult to derive a process-rating component using the original definitions (purpose, activities and tasks), in other words the 1995 version of the ISO/IEC 12207 standard was incompatible with the requirements for process assessment.

The SPICE project working group developed the process dimension in ISO/IEC TR 15504-2 based upon ISO/IEC 12207 and this has subsequently been used as feedback into the update of ISO/IEC 12207. The ISO/IEC 12207:1995/Amd.1: 2002 resolves the assessment granularity issue and provides process purpose and outcomes to establish a Process Reference Model in accordance with the requirements of ISO/IEC 15504-2 (Annex F). The updated Process Reference Model provides definitions of processes in terms of process purpose and outcomes, together with an architecture describing relationships between the processes.

The other changes in the amendment are also significant. There are 6 completely new processes consisting of: User Support, Product Evaluation, Usability, Asset Management, Reuse Programme Management and Domain Engineering.

The management process now consists of 6 sub-processes: Organizational Alignment, Organization Management, Project Management, Quality Management, Risk Management and Measurement, and Training is now greatly expanded to cover Human Resource Management (and also covers knowledge management). In addition to changes promoted by ISO/IEC 15504-2, it also incorporates suggested change aspects from ISO/IEC 14598:1998 Software Engineering – Product Evaluation [6] and ISO/IEC 15939 Software Engineering – Software Process Measurement [7].

The organization must define their processes to conform to the requirements of this standard if they desire compliance to the standard.

The ISO/IEC 12207 process architecture defines the hierarchical relationship among processes, activities and tasks and the invocation rules for the software life cycle processes. The standard does not define how, or in what order, the elements of the purpose statements are to be achieved. These are defined by the organization when it specifies its detailed practices and work products. Once defined, these elements will demonstrate the achievement of process purpose when implemented.

In amendment 1 of the standard, Annex F defines a Process Reference Model. It provides descriptions of process purpose and outcomes. The overall model consists of both process level and activity level detail, as shown in the following figure [25].

In amendment 2 of the standard [26], the Supply process now has 4 component processes, and Change Request Management has been added as a support process [8].

---

[25] This diagram was created by the author as no diagram was issued with ISO/IEC 12207 Amendment 1 nor Amendment 2.

[26] Amendment 2 is being balloted in 2004. It is hoped that the standard will be completely re-issued rather than just amended in its next iteration.

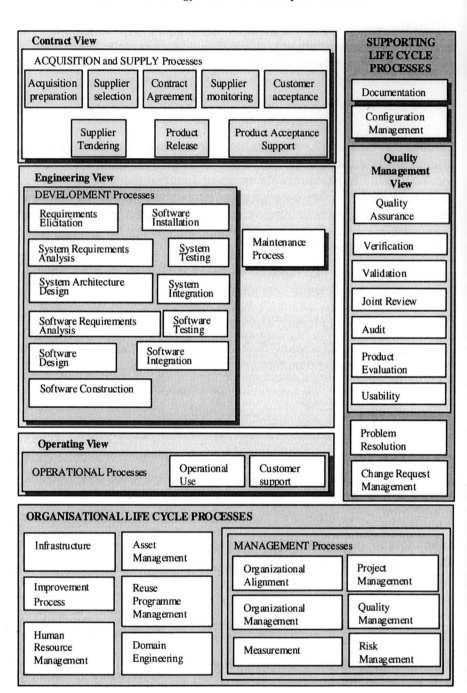

ISO/IEC 12207 Amd 2: 2004 Process Reference Model.

The organization must define its own set of applicable processes, taking into account the proposed set of processes described in Annex F of ISO/IEC 12207, its business needs (for example: is the focus on software development or use?), its own environmental parameters (for example, is it a small or large organization?) and the domain of application (for example, safety critical, automotive, space software, or computer games). In addition, the organization may define novel or alternative processes that achieve the same goals of those described in this standard, but have different detailed provisions to those prescribed in the body of this standard including additional outcomes. The tailoring may also eliminate processes that are irrelevant to the organization. This tailoring of the standard is explicitly mentioned as part of adoption of the standard.

The resultant organization's set of processes is called their standard process set. This standard process set may be further tailored in use to suit customer requirements; normally this would be done on a project-by-project basis as required.

Annex F allows the organizations own set of processes to be assessed for conformance to the ISO/IEC 12207 standard. The process definitions are evaluated against the statements of Purpose and Outcomes in Annex F for conformance to Process Reference Model requirements. To claim conformance, the organization needs to demonstrate that the implementation of the processes results in the realization of the corresponding Purpose and Outcomes provided in Annex F.

For government and industry associations, the creation of a standard process set to suit their particular domain is attractive; an example would be the V-Model. Single organizations (for example a large system and software developer) may also create conformant process sets for their own business purposes, but are less likely to explicitly validate conformance to the standard. Organizations will find that if they do create a process set based upon the standard, it will ease the process mapping needed before process assessment.

Annex F is also the applicable section for an organization that is assessing its processes in order to determine the capability of these processes. It permits assessment of the effectiveness of the processes in ways other than simple conformity evaluation, even when the organization defines novel or alternate process definitions to those in the main text of the standard.

Annex F can also be used to develop assessment model(s) for assessing processes using ISO/IEC 15504-2. The purpose and outcomes are indicators that demonstrate whether the organization's processes are being achieved.

These indicators are useful to process assessors to determine the capability of the organization's implemented process and to provide source material to plan organizational process improvement.

ISO/IEC 12207 Process Reference Model process descriptions

The ISO/IEC 12207 processes are summarized in this section in terms of their process purpose and in some cases their outcomes when successfully implemented. I will describe the Contract and the Engineering processes in a little more detail as these are the focus of the standard (they are grouped in the Primary Life Cycle process category of the standard). I have added some additional comments and explanations. Both process implementers and process assessors should become familiar with these process descriptions.

Please refer to the standard for a full description and guidance.

ISO/IEC 12207 [9] uses the following definitions:

Process Purpose: The high level objective of performing the process and the likely outcomes of effective implementation of the process. The implementation of the process should provide tangible benefits to the stakeholders.

Process Outcome: an observable result of the successful achievement of the process purpose. This includes: production of an artefact, a significant change in state and/or meeting of specified constraints, e.g. requirements, goals, etc.

Primary Life Cycle Processes

Primary Life Cycle Processes consist of Acquisition, Supply, Operation, Development and Maintenance processes. They are the main focus of the ISO/IEC 12207 standard and consequently developed in more detail, better documented and therefore easier to understand.

The standard describes a series of views within the process lifecycles. There are two views within the primary lifecycle, the contract view, and the engineering view.

The contract view consists of Acquisition and Supply processes.

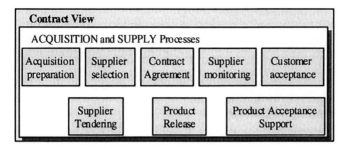

The **Acquisition Process** consists of preparation, supplier selection and monitoring, contract agreement and customer acceptance sub-processes. The purpose of the Acquisition Process is to obtain the product and/or service that satisfy the need expressed by the customer (Note: the standard assumes this is a software oriented product or service – this is apparent in the supply process).

The process begins with the identification of a customer need and ends with the acceptance of the product and/or service needed by the customer. An acquiring organization should specify (and assess) these processes. The outcomes include:

- The acquisition needs, goals, product and/or service acceptance criteria and acquisition strategies are defined and validated (including the concept or the need for the acquisition/development/enhancement; the customer's known requirements; an acquisition strategy; contractual acquisition requirements; and supplier selection criteria).

- A supplier is chosen and a clear agreement developed that expresses the expectation, responsibilities and liabilities of both the customer and the supplier. This is based on the supplier selection criteria, the supplier's proposals, process capabilities, and other contract/customer specific factors. The result is in a negotiated agreement between the customer and the supplier.

- The acquisition is monitored so that specified activities are performed and constraints such as cost, schedule and quality are met (including joint activities between the customer and the supplier; exchange of regular information on technical progress; performance against the agreed requirements; and negotiation of needed/desired agreement changes).

- An agreement between the customer and the supplier for developing, maintaining, operating, packaging, delivering, and installing the product and/or service (Contract Agreement component process).

- The customer acquires products and/or services from the supplier that satisfy their stated needs. The customer acceptance is based upon previously agreed acceptance criteria and the requirements of the agreement.
- Any identified open items have a satisfactory conclusion as agreed to by the customer and the supplier.

The **Supply Process** is the corollary to the Acquisition process. The purpose of the Supply process is to provide a product or service to the customer that meets the agreed requirements. It is therefore of more interest to suppliers, but acquiring organizations may be interested in how these processes match their acquisition processes, especially when the relationship is a close and/or long term one (for example, in partnering, just in time or Keiretsu relationships). The outcomes include:

- A response to customer's request (Supplier Tendering component process[27]).
- Supply of products and/or services that meet the agreed requirements (Product release component process).
- The products and/or services are delivered to the customer in accordance with the agreed requirements (Product Acceptance Support component process).

The standard in Annex H of Amendment 1 also describes an alternative acquisition process chain, consisting of 17 sub-processes. This process chain would be suitable for organizations that have a significant expenditure in acquisition, where they do not develop products themselves, or where acquired products are a substantial component of the total effort/cost of products (and services) they provide.

The alternate acquisition process chain starts from a policy and strategy level and progresses through proposal request and evaluation, contract agreement, acceptance and contract closure. It encapsulates an acquirer's viewpoint of system/software products, including technical and non-technical aspects such as benefits analysis, legal requirements and financial management. It also depicts the organization within a larger organizational chain by means of sub-processes such as user relationships and supplier relationships.

Annex H is offered as an alternative to the Acquisition process in Annex F. Although not explicitly stated, readers can decide whether to use all or only

---

[27] Amendment 1 had only the basic process, but Amendment 2 has the specified four equivalent component processes.

some of the sub-process descriptions from Annex H together with parts of Annex F.

The Engineering view consists of both development and maintenance processes.

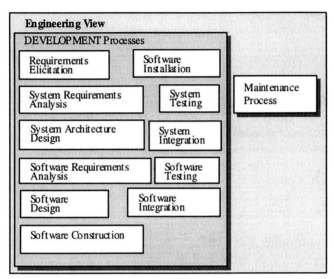

The **Development Processes** are described in detail, as they are a primary focus process area in the standard. The Development Process has 11 sub-processes. It is of primary interest to organizations that develop systems containing software (including firmware). It will be of interest to customers (especially technical customers such as systems integrators) that require formal control and monitoring of the supplier development processes. For example, when system integrators specifically order newly developed software-based products or where collaboration, subcontracting of software development or concurrent engineering is used.

The purpose of the Development Processes are to transform a set of requirements into a software product or software based system that meets the customer's stated needs. The activities of the Development Process are composed for both the Systems Developer role and the Software Developer role. The outcomes include the gathering of development requirements (at software and system levels); development of a software product or software-based system together with intermediate products that demonstrate that the end product is based upon the requirements; a way to ensure consistency is established between the products of the development process; a means to optimise system quality factors against system requirements, e.g., speed, development cost, usability, etc.; creation of evidence (for example, testing evidence) that demonstrates that the end product meets the requirements; culminating in the end product being installed according to the agreed requirements.

Each of the Development Process area sub-processes are described below.

Note that the order of description corresponds to a phased approach but this is only for convenience of reading, the reader should not assume that a phased approach is mandated.

There are many possible ways to implement the processes/sub-processes. Some possible approaches include an iterative approach where several processes are iterated before another group are performed, a cyclic approach, a partial parallel approach, as well as the phased approach. The specific approach is outside the scope of the standard as it describes 'how' the processes are implemented.

It is also not mandatory that every sub-process must be performed. For example, a software development organization may be given a complete software requirements specification and start their activities in the Software Design sub-process. Alternately, the organization may deliver software to another organization for system testing.

The **Requirements Elicitation** sub-process specifies the interface between the customer and the supplier (how they interact to determine the requirements)[28]. The purpose of requirements elicitation is to gather, process, and track evolving customer needs and requirements throughout the life of the product and/or service so as to establish a requirements baseline that serves as the basis for defining the needed work products. Requirements elicitation may be performed by the acquirer or the developer of the system, and therefore is of interest to both organizations.

The outcomes include ongoing and effective communication between the customer and supplier. The supplier agrees requirements with the customer, and defines and baselines them. A change mechanism is established and operates to evaluate changing customer needs/requirements and incorporate agreed changes into the baselined requirements. A mechanism is established for continuous monitoring of the customer needs together with a mechanism that allows customers to easily determine the status and disposition of their

---

[28] Note: In some process assessment models (including earlier versions of ISO/IEC TR 15504-5) , requirements elicitation may appear as a customer-supplier process.

requests. There is identification and management of technological and customer desired enhancements.

The **System Requirements Analysis** sub-process is focused on creation of system level requirements. The purpose of system requirements analysis is to transform the defined stakeholder requirements into a set of desired system technical requirements that will guide the design of the system. Therefore, this sub-process is of most importance to the organization that creates the system level requirements, but often of importance to the technical organization of acquirer organizations (especially when complex systems are to be developed).

The outcomes include a defined (and baselined) set of system functional and non-functional requirements describing the problem, and appropriate techniques to optimise the preferred project solution. The system requirements are analysed for correctness and testability. The system requirements' impacts on the operating environment are understood. The requirements are prioritised, approved and updated as needed, using consistency mechanisms and providing traceability between the system requirements and the customer's requirements baseline. The system requirements are communicated to all affected parties. There is an evaluation of cost, schedule and technical impact of proposed and implemented changes to the baseline.

The **System Architectural Design** sub-process converts or places requirements into a system architecture. The purpose of system architectural design is to identify which system requirements should be allocated to which elements of the system.

The outcomes include a defined system architecture design that identifies the elements of the system and meets the defined requirements, comprising the system's functional and non-functional requirements. These functional and non-functional requirements are handled in the architectural design and are allocated to the system architecture elements together with their defined internal and external interfaces. The system requirements are verified to be correctly incorporated in the system architecture. The requirements allocated to the system elements and their interfaces are traceable back to the customer's requirements baseline, and consistency and traceability between the system requirements and system architecture design is maintained (for example, when changes occur). The system requirements, the system architecture design, and their relationships are baselined and communicated to all affected parties.

Note: where a system architecture uses a product line or product family approach, this process will also consider existing system elements and decide upon reuse or modification of existing elements, or new design.

The **Software Requirements Analysis** sub-process is the first process to focus on specifically on software[29]. The purpose of software requirements analysis is to establish the requirements of the software elements of the system. In some developments, this will be performed by a different organization to the system developer. If this is the case, it becomes an important interface process between the system developer and the software developer. Normally personnel involved in the system requirement process will also have some oversight or review of the performance of the software requirements process to ensure that the system requirements have been properly translated into software requirements.

The outcomes include that the (system and customer) requirements are allocated to the software elements of the system and their interfaces are defined. The software requirements are analysed for correctness and testability. The software requirements' impacts on the operating environment are understood. A mechanism to ensure consistency and traceability is established between the software requirements and system requirements. There is a prioritisation of the software requirements' implementation. There is a mechanism to ensure that the software requirements are approved and updated (as needed). The software requirements are baselined and communicated to all affected parties. There is a mechanism to ensure evaluation of cost, schedule and technical impact of changes to the software requirements.

The **Software Design** sub-process is the first process to focus on the software product rather than requirements for software. The purpose of software design is to provide a design for the software that implements and can be verified against the requirements. In some organizations, this may be a dedicated team or designer, in others (e.g. in agile methods) it may be performed by any member of the development team.

The outcomes include a baselined software architectural design describing the software elements that will implement the software requirements, together with internal and external interfaces of each software elements. There is a detailed design describing the software units that can be built and

---

[29] Even though the standard is titled "Information Technology – Software Lifecycle Processes", as the reader will have noted, there are many processes of a generic nature (management, support) and system development nature that are relevant to software.

tested. There is a mechanism to ensure that consistency and traceability are established between software requirements and software design.

The **Software Construction** sub-process is where the fundamental software products are created. The purpose of software construction is to produce executable software units that properly reflect the software design. In large development projects, there will probably be several teams each producing one or more sets of software units.

The outcomes of Software Construction include the production of software units (defined by the design) and the definition of the verification criteria for these software units against their requirements. There is a mechanism to ensure consistency and traceability is established between software requirements, the design and the software units. The verification of the software units against the requirements and the design is accomplished.

The **Software Integration** sub-process brings the diverse outputs of the software construction process together. The purpose of software integration is to combine the software units, producing integrated software items, consistent with the software design. The software integration ensures that integrated software components demonstrate that the functional and non-functional software requirements are satisfied on either an equivalent or a complete operational platform. This may be performed by a dedicated integration team or by co-option or co-operation of the software construction teams.

The outcomes include an integration strategy for software units, consistent with the software design and the prioritised software requirements (for example, iterative software builds would incrementally meet a set of prioritised requirements and would often be iteratively integrated). The verification criteria for software items are developed to ensure compliance with the software requirements allocated to the items, and software items are verified using the defined criteria. The software items (defined by the integration strategy) are produced, tested and results of integration testing are recorded. There is a mechanism to ensure consistency and traceability is established between software design and software items. There is a regression strategy developed and applied for re-verifying software items when a change in software units (including associated requirements, design and code) occur.

The **Software Testing** sub-process is concerned with formal conformance of the software produced. The purpose of Software testing is to confirm that the integrated software product meets its defined requirements.

The outcomes include the development of criteria for the integrated software that demonstrate compliance with the software requirements. The integrated software is verified using the defined criteria and the test results are recorded. There is a regression strategy developed and applied for re-testing the integrated software when a change in software items is made.

The **System Integration** sub-process is the interface between the developed software (and the organizations producing it) and the other system components (and their involved organizations). The purpose of system integration is to integrate the system elements (including software items, hardware items, manual operations, and other systems, as necessary) to produce a complete system that will satisfy the system design and the customers' expectations expressed in the system requirements.

The outcomes of system integration include a strategy to integrate the system according to the system requirements priorities. The system integration criteria are developed to verify compliance of the integrated system elements with the system requirements (as allocated to the system elements including the interfaces between system elements), the system integration is verified using the defined criteria and the results recorded. There is a regression strategy developed and applied for re-testing the system when changes are made. There is a mechanism to ensure that consistency and traceability are established between the system design and the integrated system elements. The integrated system is constructed and it demonstrates compliance with the system design and validates that a complete set of useable deliverable system elements exists.

In system integration, it is common to use the target environment but an equivalent environment may be an alternative if it closely resembles the target environment. The decision should be based in part upon availability of the target environment and use a risk assessment if the target environment is unavailable.

The **System Testing** sub-process completes the formal aspects of system development and test. The purpose of systems testing is to ensure that the implementation of each system requirement is tested for compliance and that the system is ready for delivery.

The outcomes include the criteria for the integrated system that demonstrate compliance with system requirements. The integrated system is verified using the defined criteria, and test results are recorded. There is a regression strategy developed and applied for re-testing the integrated system when changes are made.

The **Software Installation** sub-process completes the development process life cycle. The purpose of Software installation is to install the software product that meets the agreed requirements in the target environment.

The outcomes include a software installation strategy, criteria for software installation are developed that demonstrate compliance with the software installation requirements, and the software product is installed in the target environment. There is assurance provided that the software product is ready for use in its intended environment.

Some additional notes the reader should consider about the development process.

Each of the sub-process in the development process may have one or more outputs that achieve the defined outcomes. This is defined separately by the organization implementing the sub-process.

The sub-processes can be viewed as a process chain (as described in chapter 1). In the process chain, the outputs of one process can become inputs to the next process in the process chain. However, they could also be feedback to the sub-process itself or even be inputs to an earlier sub-process in the process chain (for example in a cyclic or iterative approach). It is partially for this reason that the standard does not specify the actual lifecycle, as it is possible to implement the sub-processes in a wide variety of ways and to use different feedback loops. The way that the processes/process chain is implemented depends upon the organization's business needs, the products it produces and the people involved.

Therefore, for process assessment purposes, we only need to know the purpose and outcomes from ISO/IEC 12207.

The other process in the engineering process view is the **Maintenance Process.**

It is normally associated with the development process in the standard as part of the engineering view due to its use or interaction with many of the development sub-processes.

The purpose of the maintenance process is to modify a system/software product after delivery to correct faults, improve performance or other attributes, or to adapt to a changed environment while preserving the integrity of organizational operations. This objective may include retirement of existing system/software products. The outcomes include:

- A maintenance strategy is developed to manage modification, migration and retirement of products according to the (system/software) release strategy. (Note: this may require the creation of a release strategy that would normally be a supplier responsibility unless the system and software development is a shared customer/supplier responsibility).

- The impact of changes to the existing system on organization, operations or interfaces are identified and managed.
- The modified products are developed with associated tests or other verification/ validation methods that demonstrate that requirements are not compromised, and the affected system/software documentation is updated as needed.
- The product upgrades are migrated to the customer's environment, and products are retired from use upon request, in a controlled manner that minimizes disturbance to the customers.
- The system/software modification is communicated to all affected parties.

The Operating view consists of the operational process that takes the outputs from the engineering development and maintenance processes and ensures that there is a smooth transition of products and services into customer operation.

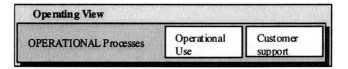

The **Operation Process** has 2 sub-processes: Operational Use and Customer Support. The purpose of the overall Operation Process is to operate the software product in its intended environment and to provide support to the customers of the software product. The overall outcomes include identification and evaluation of the conditions for correct operation of the software in its intended environment, operation of the software, and assistance and consultation to the customers of the software product in accordance with the agreement.

The purpose of **Operational Use** is to ensure the correct and efficient operation of the product for the duration of its intended usage and in its installed environment. The outcomes include identification of the operational risks for the product introduction and operation and monitoring and management of the operational risks. The criteria for the operational use are developed that demonstrate compliance with the agreed requirements, and the product is operated in its intended environment according to requirements and criteria.

The purpose of **Customer support** is to establish and maintain an acceptable level of service through assistance and consultation to the customer to support effective use of the product. The outcomes include identification of the service needs for customer support and monitoring on an ongoing basis. The customer satisfaction is evaluated on an ongoing basis for both the support services being provided and for the product itself. The initiation of operational support to handle customer inquiries and requests, to resolve operational problems, and the customer support needs are met through delivery of appropriate services.

It should be noted that both the operational use and customer support processes have important dependencies and relationships to the contract view (acquisition/supply) and the engineering view (development/maintenance) processes.

Operational use often follows the software installation sub-process (development). However, it should have provided earlier some user environment requirements at the start of development (either through the requirements elicitation process or through the acquisition preparation sub-process). It may also interact with the customer acceptance sub-process in the acquisition process chain when systems/software are acquired rather than developed.

In a similar manner, customer support interacts with the maintenance process (and consequently the development process if software maintenance is performed within the organization). An effective customer support process provides not only support to the customer, but also a means for the software product developer/maintainer to determine what changes and problem solutions are required. For problems, the customer support process may make use of the problem resolution process (in supporting life cycle processes), or use its own process.

Together with operational use, customer support may provide an important source of input for future product development, both for new versions of existing products and for new products.

It is for this reason that the operational use and customer support processes are grouped into the primary lifecycle processes.

Supporting Life Cycle Processes

The second major grouping of processes is the **Supporting Life Cycle Processes.** There are 10 Supporting Life Cycle processes. In general, the supporting life cycle processes can be applied to support the primary and organizational life cycle processes.

Four processes are generic: Documentation, Configuration Management and Problem Resolution. They are generally applicable to the primary and organization processes and outcomes. The other seven (7) processes are focused on supporting quality management[30] of the other processes.

The **Documentation Process** is a generic process. The purpose of the Documentation process is to develop and maintain the recorded software information produced by a process.

The **Configuration Management Process** is a generic process. The purpose of the Configuration management process is to establish and maintain the integrity of all the work products of a process or project and makes them available to concerned parties.

The **Problem Resolution Process** is a generic process. The purpose of the Problem resolution process is to ensure that all discovered problems are analysed and resolved and that trends are recognized.

The **Change Request Management Process** is a generic process. The purpose is to ensure that change requests are managed, tracked and controlled. Previously in Amendment 1 there was no separate process and the problem resolution process was referred to instead. This process now allows non-problem related changes.

The following processes are part of the quality management view of processes.

The purpose of the **Quality Assurance process** is to provide assurance that work products and processes comply with predefined provisions and plans.

The purpose of the **Verification process** is to confirm that each software work product and/or service of a process or project properly reflects the specified requirements.

---

[30] ISO/IEC 12207 uses the term: Quality Management view, which is focused on application within a project or development programme. There is also the Quality Management process, which is focused on the overall management/organization level.

The purpose of the **Validation process** is to confirm that the requirements for a specific intended use of the software work product are fulfilled.

The purpose of the **Joint Review process** is to maintain a common understanding with the stakeholders of the progress against the objectives of the agreement and what should be done to help ensure development of a product that satisfies the stakeholders. Joint reviews are at both project management and technical levels and are held throughout the life of the project.

The purpose of the **Audit process** is to independently determine compliance of selected products and processes with the requirements, plans and agreement, as appropriate.

The purpose of the **Product Evaluation process** is to ensure through systematic examination and measurement that a product meets the stated and implied needs[31] of the users of that product.

The purpose of the **Usability process** is to ensure the consideration of the interests and needs stakeholders in order to enable optimising support and training, increased productivity and quality of work, improved human working conditions and reducing the chance of user rejection of the system.

The supporting life cycle processes were initially devised as a group to support the primary life cycle processes, especially the development process. Some of the process descriptions still reflect this relationship (for example verification focuses on the software product). However, processes such as joint review, product evaluation and usability are also highly applicable to the acquisition process and the operational processes.

The audit process is not only applicable to the primary life cycle processes, but to any process in the model, and provides an important input to the Quality Management process in the organizational life cycle.

Organizational Life Cycle Processes

The Organizational life cycle consists of a set of management processes and a set of organizational processes. These processes are important to establish and maintain the organization's primary life cycle processes as suited to their own business needs. Organizations that have effective organizational and

---

[31] See ISO/IEC 14598, Software product evaluation for requirements for performing product evaluations.

management processes are likely to be more efficient and effective in their other processes and in establishing projects.

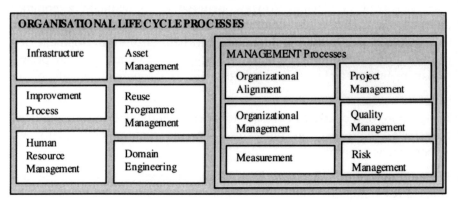

The purpose of the **Management processes** is to organize, monitor, and control the initiation and performance of any processes to achieve their goals in accord with the business goals of the organization. The Management process is established by an organization to ensure the consistent application of practices for use by the organization and the projects. While these practices are inherent to the management of an organization, they are intended to be instantiated for use by each of the organization's projects. The overall outcomes of the six management processes include:

- The definition of the scope of the process to be managed, and the identification of the activities and tasks to achieve the purpose of the process.
- The evaluation of the feasibility of achieving process goals with available resources and within existing constraints.
- The establishment of the resources and infrastructure required to perform the identified activities and tasks, the implementation of the activities and monitoring of performance.
- The review of the work products resulting from the process activities and analysis and evaluation of the results.
- Action is taken to modify the performance of the process when performance deviates from the identified activities and tasks or fails to achieve their goals.
- The demonstration of successful achievement of the purpose of the process.

There are 6 management sub-processes, which describe management in greater detail: Organizational Alignment, Organizational Management,

Project Management, Quality Management, Risk Management and Measurement.

The purpose of the **Organizational Alignment** sub-process is to enable the software processes needed by the organization to provided software products and services, to be consistent with its business goals.

The purpose of the **Organization Management** sub-process is to establish and perform software management practices, during the performance of the processes needed for providing software products and services that are consistent with the business goals of the organization.

The purpose of the **Project Management** sub-process is to identify, establish, co-ordinate, and monitor the activities, tasks, and resources necessary for a project to produce a product and/or service, in the context of the project's requirements and constraints.

The purpose of **Quality Management** sub-process is to achieve customer satisfaction by monitoring the quality of the products and services, at both the organizational and project level, to ensure they meet customer requirements.

The purpose of **Risk Management** sub-process is to identify, manage and mitigate the risks continuously, at both the organizational and project level.

The purpose of **Measurement** sub-process is to collect and analyse data relating to the products developed and processes implemented within the organization and its projects, to support effective management of the processes and to objectively demonstrate the quality of the products.

There are 6 additional organizational processes. These processes should apply across the entire organization, although it is possible for the organization to apply them on a Business unit or departmental basis (especially when the organization is large and complex).

The purpose of the **Infrastructure** process is to maintain a stable and reliable infrastructure that is needed to support the performance of any other process. The infrastructure may include hardware, software, methods, tools, techniques, standards, and facilities for development, operation, or maintenance.

The purpose of the **Improvement** process is to establish, assess, measure, control, and improve a software life cycle process.

The purpose of **Process Establishment** process is to establish a suite of organizational processes for all life cycle processes as they apply to its business activities.

The purpose of **Process Assessment** process is to determine the extent to which the organization's standard processes contribute to the achievement of its business goals and to help the organization focus on the need for continuous process improvement.

The purpose of **Process Improvement** process is to continually improve the organization's effectiveness and efficiency through the processes used and aligned with the business need.

The purpose of the **Asset Management** process is to manage the life of reusable assets from conception to retirement.

The purpose of the **Reuse Program Management** process is to plan, establish, manage, control, and monitor an organization's reuse program and to systematically exploit reuse opportunities.

The purpose of the **Domain Engineering** process is to develop and maintain domain models, domain architectures and assets for the domain.

The purpose of the **Human Resource** process is to provide the organization adequate human resources and maintaining their competencies, consistent with business needs.

- The purpose of the **Human Resource Management** sub-process is to provide the organization and projects with individuals who possess skills and knowledge to perform their roles effectively and to work together as a cohesive group.
- The purpose of **Training** sub-process is to provide the organization and project with individuals who possess the needed skills and knowledge to perform their roles effectively
- The purpose of **Knowledge Management** sub-process is to ensure that individual knowledge; information and skills are collected, shared, reused and improved throughout the organization.

ISO/IEC 12207 Amd 1:2002 has labelled some processes as sub-processes (for example, the 6 management sub-processes) while others are labelled as processes (the 6 organizational processes). In addition, it labels processes as basic, component, extended or new processes, depending upon their relationship to the process descriptions in the 1995 version of the standard. Amendment 2 uses similar descriptions.

In general, related sub-processes are grouped under a process description (for example, management, development). When a process has sub-processes, there is a description of process purpose and outcomes at the two levels (process level and sub-process level). One advantage of this is that an organization could use the top-level process description when this is sufficiently detailed for their business purposes (for example in a small organization or where the process is relatively less critical), or could use the sub-process descriptions when needed (for example in a larger and more complex organization).

The grouping of processes must be seen as a convenience for the process implementer, rather than a strict rule. The process implementer needs to consider all the relationships between the processes (well, at least those important to the business of their organization).

Other standards have created different process groupings, for example IEEE Std 1517 [10] has a process group called Cross-Project processes that span more than one software project. In this group is Domain Engineering, which provides an ability to produce products useable in multiple projects as well as take input from multiple projects. This standard is based upon ISO/IEC 12207, and was an input to the amendment of ISO/IEC 12207. For example, Asset Management and Reuse Programme Management are based in part upon IEEE Std 1517. For more information on this standard, see Carma McClure's book: Software Reuse – A Standards-Based Guide [11].

The process assessor needs only consider that the actual processes implemented in an organization have been mapped to the process dimension as described. For process assessment purposes, Annex F provides a process purpose and outcomes for each process/sub-process so either level of description can be chosen as appropriate.

5.3   Systems Engineering – Systems Life Cycle Processes

The Systems Engineering – Systems Life Cycle Processes standard is identified as ISO/IEC 15288 and the version referred to here is ISO/IEC 15288 FDIS 15288:2002 [12].

To quote the standard itself: *"This International Standard provides a common process framework covering the life cycle of man-made systems. This life cycle spans the conception of ideas through to the retirement of a system. It provides the processes for acquiring and supplying systems. In addition, this framework provides for the <u>assessment</u> and improvement of the life cycle processes."*

The standard specifies a set of processes that provide the basis to allow an organization to create their own system life cycle model. The organization can create a total environment (with an infrastructure of methods, procedures, techniques and tools) and include the trained personnel who use this to perform and manage projects through the entire system life cycle.

The organization may also use the standard as a basis for a customer (acquirer) - supplier agreement of an appropriate system life cycle and to define the interfaces between the organizations. Alternately, the organization may decide to adopt a few of the component processes, rather than the entire life cycle.

It should also be noted that although the standard is focused on man-made systems, it is not restricted to man-made artefacts, but can also include naturally occurring entities (for example water in an ecologically suitable environmental management system, or actual air movement/turbulence inputs into an aircraft flight control system) as well as the actions of the operators or personnel involved in the life cycle. This view of what comprises a system is a more holistic approach than that described in ISO/IEC 12207 as it allows for naturally occurring entities as part of an overall system. Such natural entities may have a major impact in analysis and design of the man-made system.

The life cycle staging or phasing is not prescribed (for example it may be concurrent, iterative or recursive), nor whether the system is unique, mass-produced or customized. The standard is applicable to systems of widely varying scope, complexity, innovation, domain, life span and evolution.

When an organization applies the standard, it can choose to fully conform or partially conform. For full conformance, the organization declares the (standard's) processes for which full conformance is claimed and must be able to demonstrate that all the requirements have been met. For partial conformance, tailoring shall follow the guidelines in Annex A.

The standard recognizes and interfaces with ISO/IEC 12207[32]. It has some process overlap. For example, the Implementation process (in the Technical processes group) could explicitly use ISO/IEC 12207 for the software related processes (as well as the requirements analysis, integration and operation processes). There are a number of similar processes, for example the acquisition and supply processes, which have the same purpose but are not completely identical.

---

32 It should be noted that the ISO/IEC 15288 and ISO/IEC 12207 standards are not fully harmonized with each other. The current planning is to harmonize the standards in 5 years.

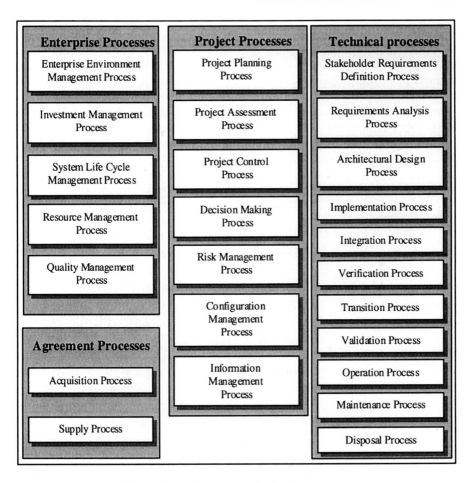

ISO/IEC 15288 System Life Cycle processes.

An organization should consider these commonalities when creating and tailoring their process framework. However, as will be described in the following section, the types of processes and the process viewpoint in ISO/IEC 15288 for many processes is quite different to ISO/IEC 12207.

The standard's processes are grouped into 4 categories: Agreement processes, Enterprise processes, Project processes and Technical processes. It is immediately apparent that the processes are grouped differently, and some have unique processes. There is a corresponding change in focus or overall purpose for each group, especially the enterprise and project process groups.

The processes may be performed in any order (or concurrently) as needed and there is no explicit hierarchy of processes.

ISO/IEC 15288 Process Reference Model process descriptions

The ISO/IEC 15288 processes are summarized in this section in terms of their process purpose and in some cases their outcomes when successfully implemented. Process implementers and process assessors should become familiar with the process descriptions.

Please refer to the standard for a full description [13].

Agreement Processes

Agreement Processes specify the requirements for the establishment of agreements with organizational entities external and internal to the organization. The Agreement Processes consist of an **Acquisition Process** – used by organizations for acquiring products or services; and a **Supply Process** – used by organizations for supplying products or services. These are similar in scope to ISO/IEC 12207 and are not described further here.

Enterprise Processes

Enterprise Processes manage the organization's capability to acquire and supply products or services through the initiation, support and control of projects. They provide resources and infrastructure necessary to support projects and ensure the satisfaction of organizational objectives and established agreements. They are not intended to be a comprehensive set of business processes that enable strategic management of the organization's business.

The purpose of the **Enterprise Environment Management Process** is to define and maintain the policies and procedures needed for the organization's business with respect to the scope of this International Standard. The outcomes include:

• Policies and procedures for the strategic management of system life cycles are provided.

• Accountability and authority for system life cycle management are defined.

• A policy for the improvement of system life cycle processes is provided.

The purpose of the **Investment Management Process** is to initiate and sustain sufficient and suitable projects in order to meet the objectives of the organization. This process commits the investment of adequate organization funding and resources, and sanctions the authorities needed to establish selected projects. It performs continued qualification of projects to confirm they justify, or can be redirected to justify, continued investment. The outcomes include:

- Investment opportunities or necessities are qualified and selected.
- Resources and budgets are identified and allocated.
- Project management accountability and authorities are defined.
- Projects meeting agreement, stakeholder and organization requirements are sustained.
- Projects not meeting agreement, stakeholder or organization requirements are redirected or terminated.

The purpose of the **System Life Cycle Processes Management Process** is to assure that effective system life cycle processes are available for use by the organization. This process provides system life cycle processes that are consistent with the organization's goals and policies, that are defined, adapted and maintained in a consistent way in order to meet the nature of individual projects, and that are capable of being applied using effective, proven methods and tools. The outcomes include:

- System life cycle processes for use by the organization are defined.
- Policy to apply system life cycle processes is defined.
- Policy to adapt system life cycle processes to meet the needs of individual projects is defined.
- Measures are defined to evaluate the application of the system life cycle processes.
- Improvements to the definition and application of system life cycle processes are undertaken.

The purpose of the **Resource Management Process** is to provide resources to projects. This process provides resources, materials and services to projects to support organization and project objectives throughout the life cycle. This includes a supply of educated, skilled and experienced personnel qualified to perform life cycle processes. This process assures that there is effective co-ordination and sharing of resources, information and technologies. The outcomes include:

- Necessary resources, materials and services are provided to projects.
- Skills of personnel are maintained or enhanced.
- Conflicts in multi-project resource demands are resolved.

The purpose of the **Quality Management Process** is to assure that products, services and implementations of life cycle processes meet enterprise quality goals and achieve customer satisfaction.

Project Processes

The **Project Processes** are used to establish and evolve project plans, to assess actual achievement and progress against the plans and to control execution of the project through to fulfilment. Individual Project Processes may be invoked at any time in the life cycle and at any level in a hierarchy of projects, as required by project plans or unforeseen events. The Project Processes are applied with a level of rigor and formality that depends on the risk and complexity of the project.

The purpose of the **Project Planning Process** is to produce and communicate effective and workable project plans. This process determines the scope of the project management and technical activities, identifies process outputs, project tasks and deliverables, establishes schedules for project task conduct, including achievement criteria, and required resources to accomplish project tasks.

The purpose of the **Project Assessment Process** is to determine the status of the project. This process evaluates, periodically and at major events, the progress and achievements against requirements, plans and overall business objectives. Information is communicated for management action when significant variances are detected.

The purpose of the **Project Control Process** is to direct project plan execution and ensure that the project performs according to plans and schedules, within projected budgets and it satisfies technical objectives. This process includes redirecting the project activities, as appropriate, to correct identified deviations and variations from other project management or technical processes. Redirection may include replanning as appropriate.

The purpose of the **Decision-making Process** is to select the most beneficial course of project action where alternatives exist. This process responds to a request for a decision encountered during the system life cycle, whatever its nature or source, in order to reach specified, desirable or optimised outcomes. Alternative actions are analysed and a course of action selected and directed. Decisions and their rationale are recorded to support future decision-making.

The purpose of the **Risk Management Process** is to reduce the effects of uncertain events that may result in changes to quality, cost, schedule or technical characteristics. This process identifies, assesses, treats and monitors risks during the entire life cycle, responding to each risk in terms of appropriate treatment or acceptance.

The purpose of the **Configuration Management Process** is to establish and maintain the integrity of all identified outputs of a project or process and make them available to concerned parties.

The purpose of the **Information Management Process** is to provide relevant, timely, complete, valid and, if required, confidential information to designated parties during and, as appropriate, after the system life cycle. This process generates, collects, transforms, retains, retrieves, disseminates and disposes of information. It manages designated information, including technical, project, enterprise, and agreement and user information.

Technical Processes

The **Technical Processes** are used to define the requirements for a system, to transform the requirements into an effective product, to permit consistent reproduction of the product where necessary, to use the product to provide the required services, to sustain the provision of those services and to dispose of the product when it is retired from service.

The Technical Processes define the activities that enable enterprise and project functions to optimise the benefits and reduce the risks that arise from technical decisions and actions. These activities enable products and services to possess the timeliness and availability, the cost effectiveness, and the functionality, reliability, maintainability, producibility, usability and other qualities required by acquiring and supplying organizations. They also enable products and services to conform to the expectations or legislated requirements of society, including health, safety, security and environmental factors.

The purpose of the **Stakeholder Requirements Definition Process** is to define the requirements for a system that can provide the services needed by users and other stakeholders in a defined environment. It identifies stakeholders, or stakeholder classes, involved with the system throughout its life cycle, and their needs and desires. It analyses and transforms these into a common set of stakeholder requirements that express the intended interaction the system will have with its operational environment and that are the reference against which each resulting operational service is validated in order to confirm that the system fulfils needs.

The outcomes include:

- The required characteristics and context of use of services are specified.
- The constraints on a system solution are defined.
- Traceability of stakeholder requirements to stakeholders and their needs is achieved.
- The basis for defining the system requirements is described.
- The basis for validating the conformance of the services is defined.
- A basis for negotiating and agreeing to supply a service or product is provided.

The purpose of the **Requirements Analysis Process** is to transform the stakeholder, requirement-driven view of desired services into a technical view of a required product that could deliver those services. This process builds a representation of a future system that will meet stakeholder requirements and that, as far as constraints permit, does not imply any specific implementation. It results in measurable system requirements that specify, from the developer's perspective, what characteristics it is to possess and with what magnitude in order to satisfy stakeholder requirements.

The purpose of the **Architectural Design Process** is to synthesize a solution that satisfies system requirements. This process encapsulates and defines areas of solution expressed as a set of separate problems of manageable, conceptual and, ultimately, realizable proportions. It identifies and explores one or more implementation strategies at a level of detail consistent with the system's technical and commercial requirements and risks. From this, an architectural design solution is defined in terms of the requirements for the set of system elements from which the system is configured. The specified requirements resulting from this process are the basis for verifying the realized system and for devising an assembly and verification strategy. The outcomes include:

- An architectural design baseline is established.
- The set of system element descriptions that satisfy the requirements for the system are specified.
- The interface requirements are incorporated into the architectural design solution.
- The traceability of architectural design to system requirements is established.
- A basis for verifying the system elements is defined.
- A basis for the integration of system elements is established.

The purpose of the **Implementation Process** is to produce a specified system element. This process transforms specified behaviour, interfaces and implementation constraints into fabrication actions that create a system element according to the practices of the selected implementation technology. The system element is constructed or adapted by processing the materials and/or information appropriate to the selected implementation technology and by employing appropriate technical specialities or disciplines. This process results in a system element that satisfies architectural design requirements through verification and stakeholder requirements through validation.

The purpose of the **Integration Process** is to assemble a system that is consistent with the architectural design. This process combines system elements to form complete or partial system configurations in order to create a product specified in the system requirements.

The purpose of the **Verification Process** is to confirm that the specified design requirements are fulfilled by the system. This process provides the information required to effect the remedial actions that correct non-conformances in the realized system or the processes that act on it.

The purpose of the **Transition Process** is to establish a capability to provide services specified by stakeholder requirements in the operational environment. This process installs a verified system, together with relevant enabling systems, e.g. operating system, support system, operator training system, user training system, as defined in agreements. The outcomes include:

- A system transition strategy is defined.
- A system is installed in its operational location.
- A system, when operated, is capable of delivering services.
- The configuration as installed is recorded.
- Corrective action reports are recorded.
- A service is sustainable by enabling systems.

The purpose of the **Validation Process** is to provide objective evidence that the services provided by a system when in use complies with stakeholders' requirements. This process performs a comparative assessment and confirms that the stakeholders' requirements are correctly defined. Where variances are identified, these are recorded and guide corrective actions. Stakeholders ratify system validation.

The purpose of the **Operation Process** is to use the system in order to deliver its services. This process assigns personnel to operate the system, and monitors the services and operator-system performance. In order to sustain services it identifies and analyses operational problems in relation to agreements, stakeholder requirements and organizational constraints. The outcomes include:

- An operation strategy is defined.
- Services that meet stakeholder requirements are delivered.
- Approved corrective action requests are satisfactorily completed.
- Stakeholder satisfaction is maintained.

The purpose of the **Maintenance Process** is to sustain the capability of the system to provide a service. This process monitors the system's capability to deliver services, records problems for analysis, takes corrective, adaptive, perfective and preventive actions and confirms restored capability.

The purpose of the **Disposal Process** is to end the existence of a system entity. This process deactivates, disassembles and removes the system and any waste products, consigning them to a final condition and returning the environment to its original or an acceptable condition. This process destroys, stores or reclaims system entities and waste products in an environmentally sound manner, in accordance with legislation, agreements, organizational constraints and stakeholder requirements. Where required, it maintains records in order that the health of operators and users, and the safety of the environment, can be monitored. The outcomes include:

- A system disposal strategy is defined.
- Disposal constraints are provided as inputs to requirements.
- The system elements are destroyed, stored, reclaimed or recycled.
- The environment is returned to its original or an agreed state.
- Records allowing knowledge retention of disposal actions and the analysis of long-term hazards are available.

As stated in the introduction to this standard, ISO/IEC 15288 takes a different systems viewpoint to ISO/IEC 12207 (which only partially addresses systems). The ISO/IEC 15288 standard is oriented towards a total lifecycle from idea formulation to retirement and disposal. When an organization defines and implements specific processes, the process implementers are recommended to consider this standard as a potential input.

Parts of the standard can be replaced by ISO/IEC 12207 (for example, some of the technical processes), but there are many processes that are unique. The enterprise process group comprises processes such as Enterprise Environment Management and Investment Management, which are beyond the scope of software lifecycle standards.

Processes such as Decision Making and Resource Management explicitly address inter-personal issues. Where these processes are considered important to an organization, it is recommended that they be investigated more closely in the standard as they provide a useful basis for an organization to implement its own processes.

In addition, when an organization needs to consider the development of systems that heavily interact with natural entities such as the external environment, the standard can provide a useful basis for definition of processes that handle this interaction.

For organizations performing internal assessments for the purpose of improvement, it is recommended that the process assessors become more familiar with this standard. Since it provides a systems view of processes and additional process descriptions, it provides potential inputs for process assessment and improvement.

5.4   Other Candidates for Process Reference Models

There are several other candidates for Process Reference Models. It is expected over time that some of these industry specific models will be developed further to better serve the particular needs of that industry. These include:

- ITIL® (Information Technology Infrastructure Library – currently considered for further development)
- OOSPICE (Component Based Development – already developed)
- S9K (SPICE for ISO 9000 – draft developed)
- Automotive SPICE (currently under development)
- V Model [14] (currently under redevelopment)

A brief overview of ITIL®, OOSPICE® and S9K are provided here.

Information Technology Infrastructure Library

The Information Technology Infrastructure Library [15] recognizes that organizations are increasingly dependent upon Information Technology (IT) to satisfy their corporate aims and meet their business needs. Information Technology Service Management is concerned with delivering and supporting Information Technology services that are appropriate to the business requirements of the organization.

ITIL® provides a comprehensive, consistent and coherent set of best practices for IT Service Management processes, promoting a quality approach to achieving business effectiveness and efficiency in the use of information systems. The best-practice processes described in the ITIL book both support and are supported by the British Standards BS 15000:2000, Specification for IT Service Management [16], and in turn underpin the ISO quality standard ISO 9000 for IT related quality management.

ITIL® processes[33]:

- Configuration Management
- Change Management
- Release Management
- Incident Management
- Problem Management
- Service Desk
- Service Level Management
- Capacity Management
- Financial Management for IT Services
- Availability Management
- IT Service Continuity Management
- Customer Relationship Management
- ICT Infrastructure Management
- Application Management
- Security Management
- Environmental infrastructure processes
- Project Management

OOSPICE

The OOSPICE project consortium gathered academic and industry partners to produce a development model and a process assessment model. The consortium includes Kepler University Linz, Computer Associates Belgium, WAVE Solutions Information Technology, Huber Computer

---

[33] Taken from the Institution's Code of Practice for IT Service Management (PD0005)

Datenverarbeitung GmbH, University of Boras, Volvo Information Technology, Griffith University and COTAR, UTS [17].

OOSPICE started looking at Object Oriented technologies and has expanded to cover Component Based Development. The purpose of component-based development is to increase the quality of products by better specifying application design, component provisioning and assembly of applications (integration centric development) [18].

OOSPICE provides extensions to the ISO/IEC 15504 Process Reference Model to cover component based software engineering.

The OOSPICE project provides a comprehensive set of documents covering:

- Component based process model
- Component based process methodology
- Process Reference Model
- Process Assessment Model and methodology
- Process Assessment tool.

In this section, I will briefly describe the Process Reference Model aspects, and in chapter 6, I will describe the Process Assessment Model.

OOSPICE groups processes into 8 groups, of which 3 groups are focused on engineering (where the most important impacts of component based development occur). The 3 groups are Modelling, Application Assembly and Component Provisioning (see the OOSPICE Process Model diagram for an overview of these processes and their component processes). In addition, there are Customer-Supplier, Support, Management, Organization and Human Resources process groups.

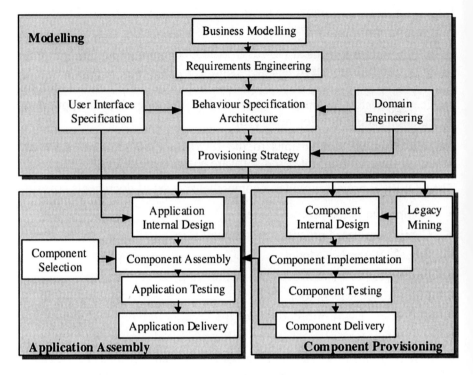

OOSPICE Engineering Processes.

The underlying architecture is based on the four-layer (metamodelling) framework of the Object Management Group [19].

Using the UML Architecture to model the OOSPICE process model in the OPEN Framework can be represented as follows.

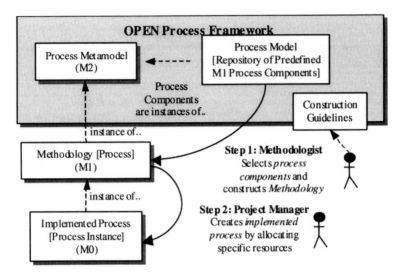

Mapping OOSPICE Process Model using OMG/UML framework.

The OOSPICE process model is similar to the process component repository in OPEN. It specifies purpose, outcomes, tasks and associated work products. The OOSPICE process model also uses the UML modelling terminology. Two main actors are defined, the Methodologist who constructs the methodology based upon use of existing process components, and the project manager who implements the methodology as a specific process instance by allocating resources to the various process components.

If we expand the above to illustrate the mapping to the UML architecture, we have the process model that comprises the set of process components specifying what needs to happen (but not *how*), and specific process component descriptions specifying how to implement the process components (tasks and techniques). These are combined with tailoring guidelines to provide the methodology. The tailoring guidelines are also used by the project manager when implementing the processes (process components).

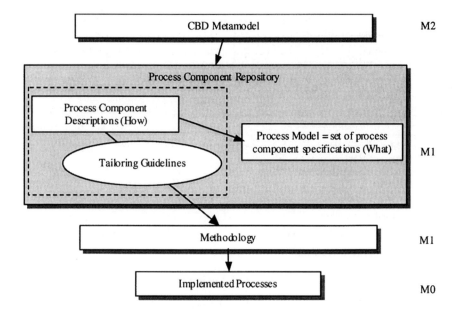

OOSPICE mapping to OMG/UML modelling.

The results is a set of process components and methodologies that can be used directly by industry for component based development and assessment.

The Process Model covers the tasks (as required to meet the Object Management Group modelling), and the purpose and outcomes needed for a Process Reference Model conformant to ISO/IEC 15504.

The OOSPICE Process Reference Model follows the ISO/IEC 15504 recommendations and specifies only the purpose and outcomes (similar to ISO/IEC 12207 Amd 1 Annex F).

For a more detailed description of the component based development models and methodologies, reader should obtain the book Business Component-Based Software Engineering [20] or refer to the OOSPICE website at www.oospice.com.

SPICE for ISO 9000

The SPICE for ISO 9000 (S9K) project has developed an ISO/IEC 15504 conformant assessment and reference model for ISO 9000 [21].

It has been derived from a European Space Agency funded project called SPICE for Quality Management in the Space Industry (S9kS) [22]. By definition, space quality assurance within European Space Agency overlaps strongly with the scope and goals of quality management addressed by the ISO 9000 family of standards and the project derived a process assessment approach covering both ISO 9000 and the space specific quality standards created by the European Cooperation for Space Standardisation.

The S9K reference model documents the set of quality management system processes that are fundamental to good quality management systems and cover practices that are required to perform these practices. The processes are described in terms of process purpose and a list of outcomes that support achievement of the process purpose.

The S9K Process Reference Model groups the Purposes and Outcomes into the six process categories of quality management system, i.e., Organizational, Customer Focus, Resource and Facilities Management, Purchasing Assurance, Core Processes (for Quality Management) and Measurement, Analysis and Improvement.

Customer Focus	Organisation
Requirements Determination	Management Direction
Customer Communication	Quality Establishment
Customer Property	Management Review
Product Acceptance & Delivery	Document Control
	Record Control
Core Processes	**Resource and Facilities Management**
Design & Development Management	Infrastructure Management
Design & Development Control	Work Environment Management
Technical Review	Human Resource Management
Production and Service Provision	Control & Monitoring of Measuring
Control	Devices
Product Preservation	
Design & Development	**Measurement, Analysis &**
ChangeManagement	**Improvement**
Product Identification	Internal Audit
	Process Monitoring and Measurement
Purchasing Assurance	Product Monitoring and Measurement
Supplier Selection	Non-conforming Product Control
Purchasing Requirements	Corrective Action
Supplier Monitoring	Preventive Action
Incoming Inspection	Continual Improvement

S9K Process Reference Model overview.

The organization must define the various detailed activities, tasks, and practices being carried out to produce work products. These performed tasks, activities, and practices, and the characteristics of the work products produced, are indicators that demonstrate whether the specific purpose is being achieved.

5.5   Summary - Process Reference Models

In this chapter, I briefly described some of the main Process Reference Models. As readers will have noticed, the models vary widely. Some of the variations are:

- Domain of the model (software, system, component based development, IT).
- Intended context of use (type and size of organization).
- Community of interest (general, space, automotive, aviation).
- Process coverage.
- How the processes are described (from textual descriptions to UML).

All the Process Reference Models have process descriptions that describe the basic process (purpose and outcomes), and each process has a unique identification. This is sufficient for ISO/IEC 15504 compliance.

The variety had both positive and negative aspects.

One negative aspect is that the implementer and assessor need to understand (or learn) different nomenclatures and terminologies to make use of some of the models.

A very positive aspect is that the different process coverage of various models provides the process implementer and process assessor with a wide variety of possible implementations and assessment process descriptions, some of which may better suit the organization's business.

Since ISO/IEC 15504 has made the Process Reference Model an external entity, there have been several communities of interest creating their own models, better suited to their industry/community. This trend is expected to continue as other communities of interest see the benefits of addressing their specific concerns. While it is not expected that there will be a huge proliferation of models, the creation of additional models will strengthen the adoption of ISO/IEC 15504 as a process assessment standard.

6 PROCESS ASSESSMENT MODELS

In this chapter, I describe conformant Process Assessment Models that are needed to perform process assessments.

I look at the ISO/IEC 15504 requirements, then at the most popular Process Assessment Model: the exemplar model in ISO/IEC TR 15504-5.

I then describe some other conformant models including SPiCE for SPACE and OOSPICE, and very briefly the current state of development of Automotive SPICE. This is followed by an overview description of the SEI CMMI® and the FAA-iCMM®.

Finally, I look at how to use the assessment indicators in these Process Assessment Models in rating the capability of processes.

This chapter will be of interest to anyone designing or assessing processes, including assessors, process owners, process experts, and developers of assessment models.

6.1 ISO/IEC 15504-2 Requirements

ISO/IEC 15504-2 requires a conformant or compliant Process Assessment Model that provides a level of detail that ensures consistent assessment results.

The Process Reference Model alone cannot be used as the basis for conducting reliable and consistent assessments of process capability since the level of detail is not sufficient.

Therefore, a Process Assessment Model is needed to form the basis for the collection of evidence and rating of process capability. The Process Assessment Model shall be based on process management principles, using the approach that the capability of a process can be assessed by demonstrating the achievement of each of its process attributes. Stated another way, this means that an overall assessment of capability can be built up from assessment of each of its components.

Any conformant Process Assessment Model must provide at least a **two-dimensional view** of process capability. In one dimension, it describes a set of process entities (process areas) that relate to the processes defined in the specific Process Reference Model (s); this is termed the **Process Dimension**.

In the other dimension, the model describes capabilities that relate to the process Capability Levels and process attributes defined in ISO/IEC 15504-2; this is termed the **Capability Dimension**.

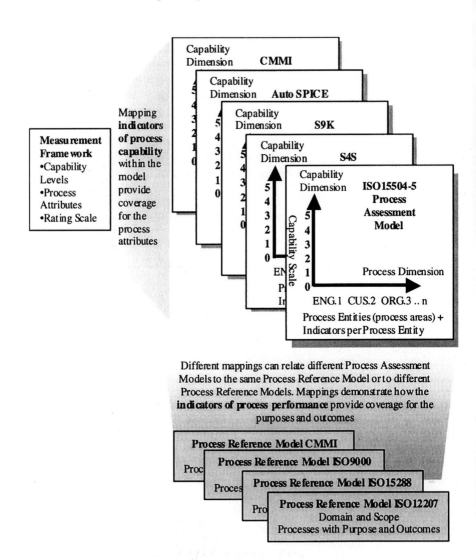

Process Assessment Model Relationships.

The diagram depicts several important aspects in the standard:

- It is possible to have many conformant Process Reference Models (horizontal axis).
- It is possible to have many conformant Process Assessment Models.
- It is possible to map one or more Process Assessment Models to one or more Process Reference Models.
- All the Process Assessment Models must map to the Measurement Framework of ISO/IEC 15504 (Capability dimension on vertical axis)

A Process Assessment Model may be related to one or more Process Reference Models. For conformance, it shall be related to (at least) one conformant Process Reference Model.

In order to ensure consistency and repeatability of assessments, a Process Assessment Model shall contain a definition of its purpose, scope, elements and indicators, mapping to the Measurement Framework and specified Process Reference Model(s), and a mechanism for consistent expression of results.

6.2 Nature and Use of indicators

> A Process Assessment Model shall be based on a set of indicators that explicitly address the purposes, as defined in the selected Process Reference Model, of all the processes within the scope of the Process Assessment Model, and that demonstrate the achievement of the process attributes within the Capability Level scope of the Process Assessment Model.
>
> The indicators focus attention on the implementation of the processes in the scope of the model.
>
> [ISO/IEC 15504-2, 6.3.4]

A model must document a set of indicators of process performance and process capability that enable judgments of process capability to be soundly based on objective evidence. The assessment indicators represent various types of objective evidence that might be found when a process is used, and therefore provide a basis to judge achievement of capability.

There is a clear expectation that the indicators will fall into two categories, related to the two dimensions of the model:

- factors that indicate the performance of the process, and
- factors that indicate its capability.

In selecting a model, clear attention should be given to the use of indicators in the model, the comprehensiveness of the indicator set, and the applicability of the indicator set.

- **Indicators of process performance** provide coverage for the purposes and outcomes. **Indicators of process capability** within the model provide coverage for the process attributes. At **Capability Level 1,** the process attribute is of process performance and is written in ISO/IEC 15504-2 in such a way as to be common to all processes. However, the evidence that demonstrates the process is performed is specific to each process (for instance an indicator of process performance for software testing is different to one for project management).

The relevant **process performance indicators** will be different from process to process but will generally consist of:

- identified work products that are input to the process (input work products);
- identified work products that are produced by the process (output work products); and
- actions taken to transform the input work products into output products (base practices).

A base practice is an activity that addresses the purpose of a particular process. When the organization consistently performs the base practices associated with a process, it will consistently achieve its process purpose. A coherent set of base practices is associated with each process in the process dimension. The base practices are described at an abstract level, identifying "what" should be done without specifying "how" (they represent the unique, functional activities of the process).

In an assessment, the output work products and actions taken (refer Base Practices) are the main indicators of process performance for Capability Level 1.

In the higher levels of the capability dimension, the **indicators of process capability** are related to the definitions of the attributes of process capability for each Capability Level. In ISO/IEC 15504-2, there are two process attributes per Capability Level for levels 2 to 5 and these work together to enhance the capability of process performance (in other words, how well the process is performed). Each process attribute describes some facet of the overall capability of managing the effectiveness of a process in order to achieve its process purpose.

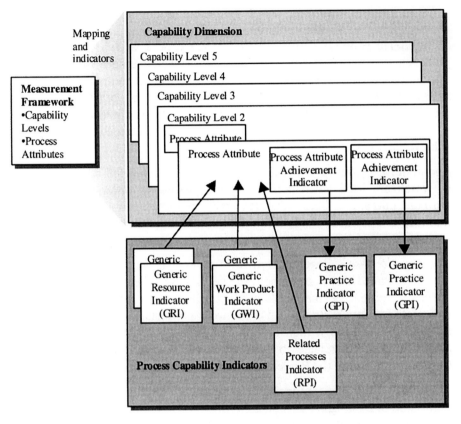

Capability Levels 2 to 5, Process Attributes and Process Capability Indicators.

The Process Attribute(s) consists of a summary, generic description of the attribute for the particular Capability Level and a list of process achievement attribute indicators. The achievement attributes describe the type of evidence that would substantiate an assessor's judgment of the extent that the attributes of process capability outcomes are achieved. These attributes are called Process Capability Indicators, and are the principal indicators of process capability.

The Process Attribute is supported by four types of Process Capability Indicators that provide evidence of process capability. In the draft version of ISO/IEC 15504-5 are called:

- Generic Practice Indicators (GPI)
- Generic Resource Indicators (GRI)
- Generic Work Product Indicators (GWI)

- Related Process Indicators (RPI)

The **generic practice indicators (GPI)** are generic activities that provide guidance on implementation that ensure the achievement of the process attribute. They reflect one facet of the achievement of the process attribute; many of them concern management practices, i.e. practices that are established to support the process performance as it is characterized at level 1. There are at least as many generic practice indicators as attribute achievement items.

The **generic resource indicators (GRI)** are the associated resources, some of which are used when performing the process in order to achieve the attribute outcomes. These resources may include human resources, tools, methods and infrastructure. They are related to the process attribute in the standard.

The **generic work-product indicator (GWI)** is a work product that is typically related to the enactment of the process. These work products are normally subsets of the work products defined as process performance indicators (at Capability Level 1). The work products are generally produced by related processes, which contribute to the achievement of the process attribute. They are related to the process attribute in the standard and the Capability Levels of the Process Assessment Model.

The **related-processes indicators (RPI)** identify processes from the process dimension of the Process Assessment Model that are linked with the attribute (they support process performance as described by the generic practice indicators). The performance of a related process supports the organizational unit to achieve the outcomes of the process attribute.

Note: In a specific Process Assessment Model, the terminology used for the four indicators may be different and may even be combined. In ISO/IEC TR 15504-5, the terminology used is different, but during its current revision, it is intended to adopt the same terminology as above.

CAUTION: A Process Assessment Model contains descriptions of processes, often grouped in process categories for convenience in *assessing* process performance. No Process Assessment Model is a complete description of all implemented business processes nor is it a specific description of any actual process implementation. Rather the model attempts to describe generically *what* a business process should achieve and provides typical *example* types of evidence. While it provides a useful supplemental source of practices and work products for a process, process implementers should take care to first ensure their processes meet organizational needs.

This may validly lead to a different composition of processes to those described in a Process Assessment Model. Consequently, assessors should take care to convert/translate the Process Assessment Model indicators against what the organization actually does.

6.3   The ISO/IEC 15504-5 Exemplar

Part 5 of ISO/IEC 15504 contains one particular example of a conformant Process Assessment Model (called an exemplar model). As I will describe later, there are other conformant Process Assessment Models that use such a different approach, but ISO/IEC 15004-5 provides a simple example and I will describe it in some detail here.

The currently published version is ISO/IEC TR 15504-5:1998 [23]. This is based upon and refers to the older Capability Levels in the 1998 version of the Technical Report. A new version is currently being prepared for balloting, planned for 2004. Some reference will be made to the new draft version [24] where compatibility with ISO/IEC 15504-2:2003 (part 2) is required, this is marked as [2003].

Part 5 is useable as a complete Process Assessment Model and in fact is the basis for many assessment models currently in use. It provides an additional level of detail required to perform reliable and consistent assessments, consisting of sets of attribute indicators for process performance and capability.

In addition, part 5 also provides an example of how to define and document a conformant assessment model that can be followed when a model developer wishes to create their own Process Assessment Model.

Part 5 uses specific terminology in relation to process assessment and some of these terms are described here to aid understanding.

- **Attribute indicator**: an assessment indicator that supports the judgment of the extent of achievement of a specific process attribute. Indicators may be related to performance of a particular process (specific practice indicators) or the capability of a process (generic practice indicators).
- **Base practice**: an activity that, when consistently performed, contributes to achieve the purpose of a particular process. In other words, a basic practice or activity, which may be a managerial, engineering or service activity.
- **Generic practice**: a management activity that addresses the implementation of a specific process attribute associated to any Capability Level from level 2 up to level 5. [2003]

- **Management practice**: a management activity or task that addresses the implementation or institutionalisation of a specific process attribute. Management practices are related to achievement in the capability dimension from level 2 up to level 5. (These are called generic practices in the new draft version [2003].)

The remainder of this chapter will describe the ISO/IEC TR 15504-5 using the existing document structure so that all components of the conformant model are covered.

Structure of the Process Assessment Model

The Process Assessment Model structure in ISO/IEC TR 15504-5 is based upon the Process Reference Model in ISO/IEC TR 15504-2 (Part 2 of the standard). Part 2 and 5 were initially based upon ISO/IEC 12207, but were modified during development to more precisely state process purpose and outcomes in a way suitable to use in assessments. This reformulation of the process descriptions has in turn been adapted in the ISO/IEC 12207 Amendment 1. In addition, part 5 contains processes (e.g. organizational processes) that were outside the scope of ISO/IEC 12207.

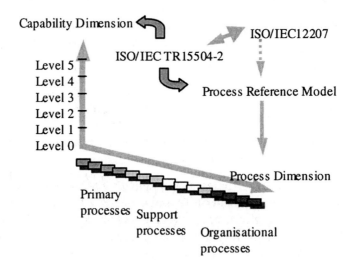

Derivation of ISO/IEC TR 15504-5:1998 Process Assessment Model.

Note: In the current part 5 (ISO/IEC TR 15504-5:1998 version), the Process Reference Model is contained in part 2 of the 1998 version of the standard. In the 2003 draft version of part 5, ISO/IEC 12207 replaces part 2 as the related Process Reference Model [2003].

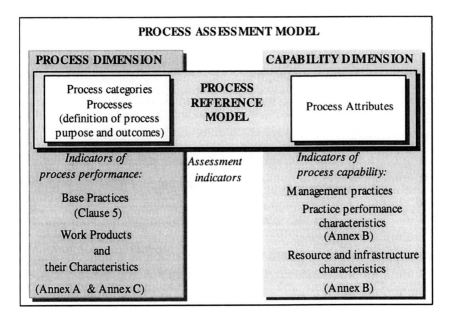

Relationship or Process Assessment and Process Reference Model Indicators.

Process Dimension

The ISO/IEC TR 15504-5 Process Assessment Model groups the related processes in the process dimension into 5 process categories.

The **Customer-Supplier** process category [acronym: **CUS**] consists of processes that directly interface with the customer, including acquisition, supply and transition of the software to the customer, and provide for the correct operation and use of the software product and/or service. Some processes have been further specified into component processes (for example CUS.1) where the additional level of detail is required.

CUS.1	**Acquisition Process**
	CUS.1.1 Acquisition Preparation Process
	CUS.1.2 Supplier Selection Process
	CUS.1.3 Supplier Monitoring Process
	CUS.1.4 Customer Acceptance Process
CUS.2	**Supply Process**
CUS.3	**Requirements Elicitation Process**

CUS.4   Operation Process

CUS.4.1 Operational Use Process

CUS.4.2 Customer Support Process

The **Engineering** process category [acronym: **ENG**] consists of processes that directly specify, develop, or maintain the software product, including its relation to the system and customer documentation.

ENG.1   Development process

ENG.1.1   System requirements analysis and design process

ENG.1.2   Software requirements analysis process

ENG.1.3   Software design process

ENG.1.4   Software construction process

ENG.1.5   Software integration process

ENG.1.6   Software testing process

ENG.1.7   System integration and testing process

ENG.2   System and software maintenance process

The **Management** process category [acronym: **MAN**] consists of processes that may be used by anyone who manages any type of project or process within a software life cycle.

MAN.1   Management Process

MAN.2   Project Management Process

MAN.3   Quality Management

MAN.4   Risk Management

The **Organization** process category [acronym: **ORG**] consists of organization-wide processes that establish the overall goals of the organization (including vision, mission and culture) and develop related assets (process, product, and resource), which help the organization achieve its goals.

ORG.1   Organizational alignment process

ORG.2   Improvement process

ORG.2.1   Process establishment process

ORG.2.2   **Process assessment process**

ORG.2.3   **Process improvement process**

ORG.3   **Human Resource Management process**

ORG.4   **Infrastructure process**

ORG.5   **Measurement Process**

ORG.6   **Reuse Process**

The **Support** process category [acronym: **SUP**] consists of processes, which may be employed, by other processes (from any process category) at various points in the software product/service life cycle.

SUP.1   **Documentation Process**

SUP.2   **Configuration Management Process**

SUP.3   **Quality Assurance Process**

SUP.4   **Verification Process**

SUP.5   **Validation Process**

SUP.6   **Joint Review Process**

SUP.7   **Audit Process**

SUP.8   **Problem Resolution Process**

Each process in the assessment model is described in terms of a process purpose statement. These statements contain the unique functional objectives of the process when instantiated (implemented) in a particular environment. There is a coherent set of base practices associated with each process. Each base practice is an activity that addresses part of the purpose of the process, and consistent performance of the base practices will help to consistently achieve the process purpose.

The base practices are described at an abstract level, identifying "*what*" should be done (the functional activities) without specifying "*how*" it should be done.

Implementing the base practices of a process is the first step in achieving process Capability Level 1. The base practices in a process produce work products that are at least marginally usable in achieving the purpose of the process. In this Process Assessment Model, each work product has a defined set of characteristics that may be used to assess the effective implementation of a process. If the base practices are performed and the work products demonstrate achievement of the process purpose, then the process will be assessed as achieving process Capability Level 1.

However, the performance of the base practices at this level does not ensure that the process is performed consistently, properly planned, predictable or result in products that necessarily meet all their requirements. The performance of the process to meet these requirements needs to meet higher Capability Level process attributes (see next section).

ISO/IEC TR 15504-5 clause 5 contains a complete description of the base practices. Annex A lists the processes and their related work products. Annex C lists the key characteristics of the work products. The following table provides an example of a process description in ISO/IEC TR 15504-5.

Example ISO/IEC TR 15504 Process Description – Quality Management

Process Description – Quality Management	Comments
MAN.3 Quality management process The purpose of the *Quality management process* is to monitor the quality of the project's products and/or services and to ensure that they satisfy the customer. The process involves establishing a focus on monitoring the quality of product and process at both the project and organizational level. As a result of successful implementation of the process: • quality goals based on the customer's stated and implicit quality requirements will be established for various checkpoints within the project's software life cycle; • an overall strategy will be developed to achieve the defined goals; • identified quality control and assurance activities will be performed and their performance confirmed; • actual performance against the quality goals will be monitored; • appropriate action will be taken when quality goals are not achieved. NOTE 1 This process supports performance of the process attributes 4.1 and 4.2 in those instances where it is invoked. NOTE 2 This process goes beyond the quality control and assurance activities performed in *Quality assurance process* (SUP.3), to provide an overall approach to meeting the stated and implicit requirements of the customer. Base Practices : **MAN.3.BP1 : Establish quality goals.** Based on the customer's stated and implicit requirements for quality, establish quality goals for the product and process that can be evaluated throughout the project, preferably in a quantitative manner. NOTE For projects involving the development of software quality goals are established for various checkpoints within the project's software life cycle. **MAN.3.BP2 : Define overall strategy.** Develop an overall strategy at the project and organizational level to achieve the defined goals by defining the metrics that will measure the results of project activities and by defining acceptance criteria that will help to assess whether the relevant quality goals have been achieved. **MAN.3.BP3 : Identify quality activities.** For each quality goal, identify quality control and assurance activities which will help achieve and monitor that quality goal, both at the project and organizational level. NOTE For projects involving the development of software these activities are integrated within the project's software life cycle, see also the SUP.3 Base Practices. **MAN.3.BP4 : Perform quality activities.** Perform the identified quality assurance and control activities and confirm their performance. **MAN.3.BP5 : Assess quality.** Throughout the project and at least at the identified checkpoints within the project's software life cycle, apply the defined quality metrics to assess whether the relevant quality goals have been achieved. **MAN.3.BP6 : Take corrective action.** When defined quality goals are not achieved, take corrective or preventive action both at the project and organizational level.	The Process Assessment Model documents the process purpose, and the process outcomes (copied from the Process Reference Model). Notes may provide guidance and also indicate relation with other processes The Base Practices that support the purpose and outcomes are next described. These assessment indicators are an additional level of detail to the Process Reference Model.

| NOTE The corrective action can involve fixing the product generated by a particular project activity or changing the planned set of activities in order to better achieve the quality goals or both. The preventive action can involve modifying product specifications or process definitions, or both, to prevent recurrence of the non-achievement.
 ISO/IEC TR 15504-5:1998 5.3.1.3 | |

Example ISO/IEC TR 15504 Work Products List

Associated Work Products List – Quality Management		Comments
Input	**Output**	Annex A provides example work products typical to the process. Normally some (but not all) work products are used, but those produced must demonstrate the achievement of the process attribute. The Process Assessment Model needs to list and describe the relevant work products.
6) Work break down structure 17) Project plan 16) Business plan 20 Progress status report/record 24) Quality statement / policy 25) Quality plan 28) Quality record 29) Assessment / audit record 30) Review strategy / plan 31) Review record 41) Field measure 52) Requirement specification (customer) 84 Problem report record	4) Job procedure / practice 5) Schedule 6) Work break down structure 12) Quality goal 17) Project plan 18) Process performance data 21) Analysis result 25) Quality strategy/plan 26) Improvement opportunity 29) Assessment audit / record 31) Review record 39) Quality measure 97) Corrective action	
ISO/IEC TR 15504-5:1998 Annex A		

Example ISO/IEC TR 15504 Work Product Description (part)

Work Product Description – Quality Management				Comments
ID	WP Class	WP type	WP Characteristics	Annex C provides example work product descriptions for each Work Product. Organizations may call these work products by different names or divide it into several products.
25	*1.4 / 2.1*	Quality strategy / plan + (16)	Objectives / goal for quality. Defines the activities tasks required to ensure quality. References related work products Method of assessment / assuring quality References any regulatory requirements, standards, customer requirements Identifies the expected quality criteria Target timeframe to achieve desired quality Tasks to be performed Resource commitments Identifies the quality criteria for work products and process tasks Approved by the quality responsible organization/function	
ISO/IEC TR 15504-5:1998 Annex C				

Part 5 of the standard also classifies work products as defined in the Work product classification.

ISO/IEC TR 15504 Work Product classification

WP Category number	WP Category	WP Classification number	WP Classification
1	ORGANIZATION:	1.1	Policy
		1.2	Procedure
		1.3	Standard
		1.4	Strategy
2	PROJECT:	2.1	Plan
		2.2	Requirement
		2.3	Design
		2.4	Implementation
		2.5	Product
		2.6	Interim deliverable
3	RECORDS:	3.1	Report
		3.2	Record
		3.3	Measure
		3.4	Data

ISO/IEC TR 15504-5:1998 Annex C

Capability Dimension

The Process Assessment Model uses the process attributes from the Process Reference Model to determine process capability. The current part 5 model uses the attributes from ISO/IEC TR 15504-2: 1998. These are revised to those in ISO/IEC 15504-2:2003 for the latest draft version of part 5. It is possible to substitute the existing process attributes with the new attributes without affecting the process dimension, and either is therefore useable for process assessment. It should be noted that the terminology and type of assessment indicators for the 2003 version are altered and should be taken into account when choosing to use this version (See Appendix for example). Assessors must state which set of attributes they are using in an assessment. When the revised part 5 is approved and issued, it is expected that assessors will change to the revised set of attributes within a defined period.

In this book I will refer to the revised set of process attributes as described in ISO/IEC 15504-2:2003.

ISO/IEC TR 15504 Capability dimension

Capability Level	Description	Process Attribute	Process Attribute Description
Level 0	Incomplete process	none	
Level 1	Performed process	PA1.1	Process performance attribute
Level 2	Managed process	PA 2.1	Performance management attribute
		PA 2.2	Work product management attribute
Level 3	Established process	PA 3.1	Process definition attribute
		PA 3.2	Process deployment attribute
Level 4	Predictable process	PA 4.1	Process measurement attribute
		PA 4.2	Process control attribute
Level 5	Optimising process	PA 5.1	Process innovation attribute
		PA 5.2	Process optimisation attribute

Each of the above process attributes describes one facet of the organizational unit's capability to manage and improve the effectiveness of a process in order to achieve its purpose and contribute to the business goals of the organization.

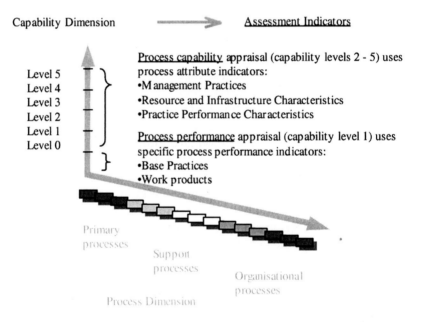

Assessment Indicators by Capability levels.

Each Capability Level represents a set of process attributes [at its own level and including those at (any) lower levels] that work together to provide a major enhancement in the capability to perform a process to achieve its purpose. In other words when assessing the achievement of a Capability

Level, it includes achievement of all lower level process attributes, the achievement is therefore cumulative. For example, to achieve Capability Level 3, the process attributes for level 2 and level 1 must be fulfilled. The Capability Levels are described in detail in chapter 2 of this book. For process improvement purposes, the levels constitute one way of progressively improving the capability of any process.

The following examples of assessment indicators come from ISO/IEC TR 15504-5:1998.

Example Capability Level 3 - Established Process

PA 3.1 – Process definition attribute

This process attribute is one of the 2 process attributes in Capability Level 3, and consists of 5 assessment indicators that in this example are management practices (MP 3.1.1 to MP 3.1.5). This first assessment indicator concerns identification of a standard process (MP 3.1.1).

Indicator Class MP 3.1.1	Indicators Identify the standard process that supports the execution of the managed process and provides documented guidance on tailoring
Practice performance characteristics	The organization's standard process documentation exists and includes: - Expected input and output work products - Work break down structure: - Tasks to be performed in line with the associated base practices - task ownership - Objective criteria for demonstrating the task completeness - Definition of internal and external interfaces, and dependencies - Quality controls: - Process entry and exit criteria - Process decision control points (possible milestones) - Expected process performance data when performing the standard process. Process performance data may address: - Resources (person, skills), - Infrastructure, duration for a task, - Cost, number of loops in a part of the process, etc. - Written organization policy for performing the process - Documentation of the standard process definition originated from the current practices performed throughout the organization exists and is validated and approved; - Standard process definition is available to all with a need-to-know in the organization: - paper documentation distributed to key process users - on-line documentation is accessible to key process users
Resource and infrastructure characteristics	Software process library Documentation tools Process modelling tools Software databases Standards Configuration management tools.
Associated processes	ORG.2.1 Process establishment SUP.1 Documentation Process

ISO/IEC TR 15504-5:1998

In addition, the model specifies additional resource and infrastructure characteristics that the assessor should look for that support management practices at Capability Level 3. Although not all the resources are mandatory, the existent of these will make the achievement of the management practice easier to achieve. Similarly, if the organization has implemented the associated processes, then the assessor is likely to find more consistent evidence that the management practice has been consistently implemented across multiple instances.

As this example demonstrates, the level of detail for the practice performance characteristics is much greater than the management practice description (and there is additional detail in the resource and infrastructure characteristics), in order to support the assessor's ability to reliably and consistently judge the achievement of the assessment indicator.

In conclusion, Part 5 of ISO/IEC TR 15504 sets out a Process Assessment Model that is compatible with the Process Reference Model in ISO/IEC TR 15504-2 by means of simple elaboration of the Process Reference Model .

The Process Assessment Model scope covers all the processes and all the Capability Levels in a one-to-one mapping of the Process Reference Model.

6.4   Additional conformant Process Assessment Models

There are several other conformant Process Assessment Models that have been or are being developed. I look in the following sections at some of these in brief.

The already developed SPiCE for SPACE and OOSPICE Process Assessment Models are conformant to the earlier ISO/IEC TR 15504-2:1998 version of the standard, while Automotive SPICE is working upon their model in parallel with the revisions to ISO/IEC 15504.

Model developers (and to a lesser extent users) need to consider changes introduced by changes in the normative part 2 of the ISO/IEC 15504 standard. In particular, they need to consider the measurement framework capability dimension aspects and the removal of a Process Reference Model in this part of the standard. They also need to consider changes introduced into the associated Process Reference Model (for example, ISO/IEC 12207).

In addition, the ISO/IEC TR 15504-5:1998 exemplar was used as a basis for some of the conformant Process Assessment Models, therefore changes introduced in the exemplar need to be considered, including terminology, layout and descriptions. See Annex 2 for the proposed changes to the exemplar model.

Please refer to chapter 3 for further requirements for conformant Process Assessment Models.

SPiCE for SPACE

SPiCE for SPACE (S4S) is a method for the assessment of software processes for the space industry. SPiCE for SPACE comprises a Process Assessment Model, an assessment method a rating process and a tool for evaluating space software processes. The method was developed by a consortium of space software suppliers and software quality experts led by Christian Völcker and Ann Cass and from SYNSPACE AG [25], under a study contract from the European Space Agency (ESA) [26] for the European space industry [27].

ESA's goals were to encourage production of the best possible software products and services; develop customer-supplier relationships based on trust, not control; promote and disseminate best practice concepts proven across the software industry; and widen their supplier base to companies traditionally outside of aerospace. The method has been in use since 2000 for assessments of space software suppliers.

SPiCE for SPACE provides both capability determination for space relevant processes and process improvement opportunity outputs. An additional advantage is that the process dimension of SPiCE for SPACE can be tailored to different classes of safety-critical software and includes an optional risk oriented dimension (Risk for Space) [28].

The Process Assessment Model adopted the same overall approach as ISO/IEC TR 15504-5, and added specific inputs from the European Cooperation for Space Standardisation (ECSS) standards [29]. The ECSS Management, Engineering and Quality standards are compatible with ISO/IEC 12207:1995 which simplified their incorporation into the SPiCE for SPACE assessment model, while allowing the dual assessment purpose to be included.

The capability dimension from the Assessment Model in ISO/IEC TR 15504-5:1998 was adopted as-is without modification or addition.

In forming the SPiCE for SPACE process dimension, all processes and base practices were adopted as-is from the ISO/IEC TR 15504-5 exemplar Assessment Model. Requirements from ECSS documents or activities from space software process models were matched with Assessment Model processes and base practices. In addition, the process dimension was augmented with space specific processes and base practices. All of the exemplar model work products are embedded in SPiCE for SPACE, as they were either used 'as is' or matched with the expected outputs of the space standard requirements. New work products and work product characteristics were created to represent the space standard outputs not covered by the ISO/IEC TR 15504-5 exemplar model. These new processes and process indicators properly incorporate space software needs into the SPiCE for SPACE Process Assessment Model.

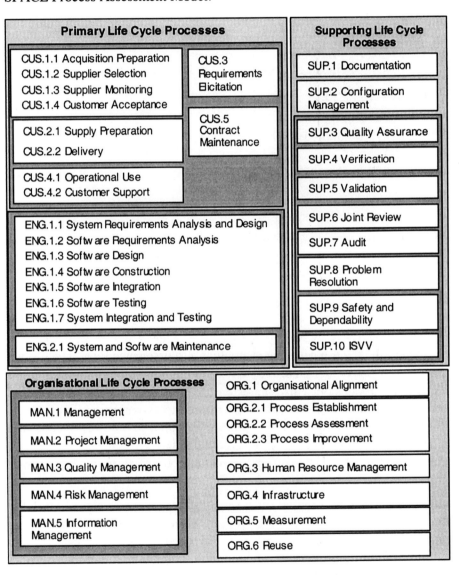

Primary Life Cycle Processes		Supporting Life Cycle Processes
CUS.1.1 Acquisition Preparation CUS.1.2 Supplier Selection CUS.1.3 Supplier Monitoring CUS.1.4 Customer Acceptance	CUS.3 Requirements Elicitation	SUP.1 Documentation
		SUP.2 Configuration Management
CUS.2.1 Supply Preparation CUS.2.2 Delivery	CUS.5 Contract Maintenance	SUP.3 Quality Assurance
		SUP.4 Verification
CUS.4.1 Operational Use CUS.4.2 Customer Support		SUP.5 Validation
		SUP.6 Joint Review
ENG.1.1 System Requirements Analysis and Design ENG.1.2 Software Requirements Analysis ENG.1.3 Software Design ENG.1.4 Software Construction ENG.1.5 Software Integration ENG.1.6 Software Testing ENG.1.7 System Integration and Testing		SUP.7 Audit
		SUP.8 Problem Resolution
		SUP.9 Safety and Dependability
ENG.2.1 System and Software Maintenance		SUP.10 ISVV

Organisational Life Cycle Processes	
	ORG.1 Organisational Alignment
MAN.1 Management	ORG.2.1 Process Establishment ORG.2.2 Process Assessment ORG.2.3 Process Improvement
MAN.2 Project Management	
MAN.3 Quality Management	ORG.3 Human Resource Management
MAN.4 Risk Management	ORG.4 Infrastructure
MAN.5 Information Management	ORG.5 Measurement
	ORG.6 Reuse

SPICE for SPACE Process Dimension.

The process dimension expands upon the process dimension in ISO/IEC TR 15504 to include Space industry related processes, base practices, work products and explanatory notes. In comparison with ISO/IEC TR 15504-5, SPiCE for SPACE has:

- 4 New Processes:
  - CUS.5   Contract Maintenance
  - MAN.5   Information Management
  - SUP.9   Safety and Dependability
  - SUP.10   ISVV
- New Component Processes
  - CUS 2.0 Supply Preparation (and Basic Supply process renamed to Delivery)
- 41 New Base Practices (altered processes shown underlined in diagram)
- 73 New Notes
- 22 New Work Products

Additional differences between ISO/IEC TR 15504-5 and SPiCE for SPACE are:

2 new categories of work products, specifically:

- ECSS Instance. This is a specific instance of an existing ISO/IEC 15504 work product, which inherits the generic characteristics of the work products but has additional characteristics added from ECSS (space standards) requirements; an example is a Maintenance Plan, which is an ECSS instance of a Plan.
- Specific Extension. This is an extension of an existing ISO/IEC 15504 work product, which has added characteristic(s) to existing ISO/IEC 15504 work products; an example is a Compliance Matrix, which is an extension of Software Product Assurance Plan.

The selection of component processes and elimination of higher-level processes:

- For the customer-supplier category, the component processes CUS.1.1, CUS.1.2, CUS.1.3, CUS.1.4 are described and CUS.1.0 is excluded.

Enhancements to SPiCE for SPACE

The following enhancements were made to the ISO/IEC TR 15504-5 Process Assessment Model in order to incorporate European space software development practices.

ISVV Process

ISVV (Independent Software Verification and Validation) is one of the organizational ways to achieve the independence of Verification and Validation. A well-defined set of standard life cycle processes is repeated by an actor completely independent from the supplier. ECSS documents require this two-fold process instantiation for highly critical software. For space projects this is applicable for software criticality class B or higher.

ISO/IEC TR 15504-5 does not describe explicitly such an ISVV instantiation of processes. However, all engineering and support activities that are repeated by ISVV activities are already defined as ISO/IEC TR 15504-5 processes: verification, validation, audit, joint review, as well as all of the engineering processes.

Only the establishment and management of an ISVV program are not addressed by ISO/IEC TR 15504-5. Thus, the new ISVV process includes Base Practices such as 'define scope of ISVV' (SUP.10.BP1) and concludes with 'report ISVV activities and results' (SUP.10.BP4). Particular work products created include ISVV plan and ISVV report. This process can be assessed even if only one of the two (independent verification or validation) is performed.

Safety Process (Planning, Analysis, Design, Verification)

In ECSS terminology, safety is a state in which the risk of harm (to persons) or damage is limited to an acceptable level [ECSS-P-001A].

ISO/IEC TR 15504-5 does not specifically address the safety of the software system. To underline the importance of the SPA requirements related to safety, a new process was incorporated into the Process Assessment Model within the SUP process category. This process covers all activities required to ensure the safety of the system.

Dependability Process (Planning, Analysis, Design, Verification)

Dependability is the degree to which the software product is available, reliable and maintainable. In the ISO/IEC TR 15504 Reference Model, no process specifically ensures software dependability. Again, in order to underline the importance of this issue for space software, a new process was added to the assessment model. The base practices are worded such that they are applicable to both the dependability and the safety process. Therefore, both processes were merged into one process, which is called "Safety and Dependability Process" (SUP.9).

Numerical Accuracy (Requirements, Design, Verification)

As an important issue to be addressed in requirements specification, design specification and testing, numerical accuracy is explicitly mentioned in notes added to the corresponding work product characteristics.

Memory Occupation and Timing (Budgeting, Design, Verification)

Memory occupation and timing are usually constraints, which are imposed through system requirements. These requirements need to be propagated and addressed through the software life cycle. Work products like Requirements Specification, Software Components Design Documents and Test Reports should document that such issues are treated carefully. Extra characteristics were added to the corresponding work products if not mentioned in the original ISO/IEC TR 15504-5 work product description.

Lessons Learned

In ISO/IEC TR 15504-5, the lessons learned process is a continuous process improvement process (ORG.2.3), which is not restricted to the end of a project. A note was added to base practice ORG.2.3.BP1 (Identify improvement opportunities), which explicitly addresses the ECSS requirement for capturing lessons learned.

In addition, a new work product, the Lessons Learned Report, was created to capture the information generated during this new practice. Its characteristics reflect ECSS-M-20 specifications. This work product becomes part of the input to many crucial software design and management processes, thus reinforcing the space project concept of continuous process improvement.

Alert System

Alerts may be raised as a reaction to non-conformance reports. The duty of the contractors is to ensure the provision of all information related to any non-conformance reported to the final customer ISO/IEC TR 15504-5 does have a Problem Resolution process, SUP.8, which provides the basis to support an alert system. A new base practice, "Prepare preliminary alert information", was added to SUP.8, to supports this practice. If an alert is broadcasted by ESA, the contractor, or its alert coordinator, must disseminate the alert within his organisation and take corrective actions as necessary. Another new base practice "Support processing of alerts" (SUP.8 BP8) describes the activities performed to respond to incoming alerts.

Subcontractor Management

In ISO/IEC TR 15504-5, all processes concerning acquisition and subcontracting are generally described. A dedicated process category (CUS) underscores the importance of this topic.

Information Management

During a space software project life cycle, every actor must be able to readily access all the information he needs in order to perform his task. Although not space-specific, the flow of information inside a project organisation is of paramount importance to success of space projects and is described by a new process, Information Management (MAN.5) in the Process Assessment Model.

Contract Maintenance

During the evolution of a software project, many circumstances may demand a revision of the business contract. In ISO/IEC TR 15504-5, no process specifically addresses contract maintenance and modification, so a new process (CUS.5) was proposed within the customer process category that admits a customer-supplier exchange and may result in a contract change.

Risk Management

Risk management is an integral part of space projects, where risks cannot be reduced to zero. Several base practices were added to MAN.4 to stress the importance of communicating and accepting risk. "Recommend acceptance", "Communicate risks" and "Accept residual risks" ensure that after risk reduction, risks are known and understood by the responsible management.

Interfaces to System Processes

In space projects, software is mostly embedded in a complex hardware system. To assure coherence between the software and other components of the system, software processes must be linked to the corresponding system processes. ISO/IEC TR 15504-5 defines such interfaces at the beginning (ENG.1.1 System Requirements Analysis and Design) and at the end of the development phase (ENG.1.7 System Integration and Testing, ENG.2 System and Software Maintenance). Space standards require further interfaces between software and system processes. For example, system PA requirements have to be considered when establishing the software PA plan, and software safety and dependability plans should be in line with the corresponding policy on the system level. In addition, milestone reviews have to be coordinated with the corresponding system reviews. SPiCE for

SPACE covers these aspects in defining appropriate input work products for the relevant processes.

Software Re-use

The re-use of software impacts both the organisation and engineering process categories. The ISO/IEC TR 15504-5 Reuse process (ORG.6) describes the production and management of software for re-use, including the definition of a re-use strategy and the establishment of a re-use infrastructure.

Regarding software engineering processes, alternative re-use solutions are analysed during the software design phase and implemented during the software construction. To reflect these activities, two new base practices, "Identify and analyse reusable components" and "Reuse software units" were added to the Software Design (ENG 1.3) and Software Construction (ENG 1.4) processes, respectively.

SYNSPACE AG continues to develop extensions and interpretative guides to SPiCE for SPACE for ESA, including a recent proposed extension to cover the Product Assessment of Software Product Reuse [30], which has been submitted for ESA approval.

OOSPICE

In this section, I will describe the Process Assessment Model of OOSPICE. Please note that I described Process Reference Model in chapter 5 [31].

OOSPICE provides a Process Assessment Model and methodology. The Process Assessment Model covers 8 groups of processes.

Customer Supplier	Support
Acquisition preparation	Maintenance
Supplier selection	Configuration Management
Supplier monitoring	Documentation
Customer acceptance	Problem resolution
Supply	Joint Review
Customer support	Verification
Operational use	Validation
	Product Evaluation
	Quality Assurance
	Audit
	Usability
Modelling	**Management**
Domain Engineering	Programme management
Business Modelling	Project management
Requirements Engineering	Risk management
Behaviour Specification Architecture	Measurement
Provisioning Strategy	Quality management
User Interface Specification	Infrastructure
Application Assembly	**Organization**
Application Internal Design	Process Establishment
Component Assembly	Process Assessment
Application testing	Process Improvement
Application delivery	Asset Management
Component selection	Reuse Programme management
Component Provisioning	**Human Resources**
Component internal design	Human resource management
Component testing	Training
Component delivery	Knowledge management
Legacy mining	

The main extensions to ISO/IEC 15504 are for the 3 engineering process areas: Modelling, Application Assembly and Component Provisioning.

The OOSPICE Process Assessment Model is integrated within the OOSPICE process model that describes the component based development methodology and the implemented processes, as well as providing the framework for the process component repository.

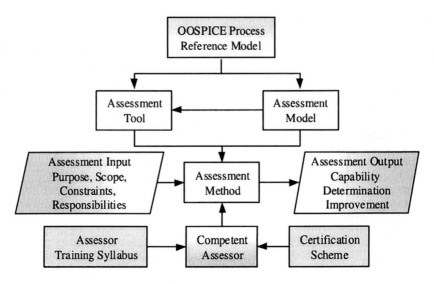

OOSPICE Assessment Methodology.

The Process Assessment Model describes for each process:

- Process Identifier and Name
- Purpose
- Outcomes
- Tasks
- Work Products

The capability dimension of ISO/IEC TR 15504:1998 has been adopted.

An assessment tool has been designed, based upon the SPICE-1-2-1 tool. For more details, see www.oospice.com.

Automotive SPICE developments

The leading automotive manufacturers around the world have increased the use of electronics both as control aids and as consumer features in the current and future development of the automobile. The use of electronic control units (ECUs) is increasingly dependent upon software for functionality and features [32]. One recent estimate of the amount of software in a modern automobile was between 40 and 90 Megabyte, distributed across multiple electronic control units, communication networks, accessories and display units.

The industry is finding that as the complexity of the ECUs and their software increases, there is a corresponding need to increase the quality of software production associated with these ECUs.

The leading automotive manufacturers in the German automotive industry through the Herstellerinitiative Software (HIS) investigated the various process assessment standards and chose to adopt ISO/IEC 15504. The group consists of Audi, BMW, Daimler Chrysler, Porsche and Volkswagen.

The Herstellerinitiative Software selected from ISO/IEC TR 15504-5 a set of basic processes for automotive assessments.

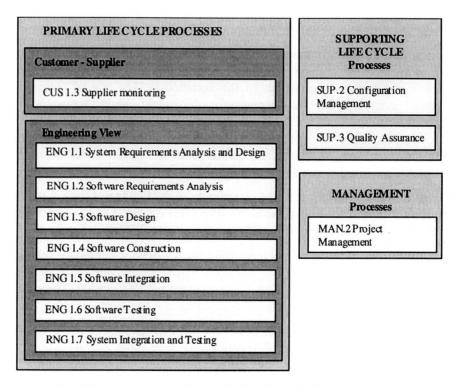

Basic subset of processes – Herstellerinitiative Software assessments.

Using this basic subset, they have assessed the relevant processes of suppliers with an aim that suppliers achieve Capability Level 2. They have set three ratings for suppliers based upon the assessments; the first is equivalent to fully achieved Capability Level 2 (> 85%), the second is above 75% achievement, and the third is anything less than this. The assessment strategy was to assess multiple potential suppliers in short assessments on a few key processes and then do a full assessment of the selected supplier for production series delivery, followed up with series delivery related measures and any improvement actions needed. The HIS also created a list of common process and resource problems in suppliers based upon these assessments as a basis for opportunities for improvement within the industry and a focus for development of Automotive SPICE.

The Herstellerinitiative Software together with other automotive manufacturers in Europe (Volvo, Peugeot, Fiat) has formed an Automotive Special Interest Group within the Procurement Forum. Experience in assessments has shown that there is a need for automotive specific guidance, so the group are creating a variant of ISO/IEC 15504 suited to their own needs called Automotive SPICE.

In general, their approach is to adopt processes from ISO/IEC 12207 Process Reference Model and adapt them to suit the automotive industry needs. They are defining a Process Reference Model and a Process Assessment Model.

The processes covered will place an additional emphasis on acquisition and supplier processes such as proposal, project requirements and contract agreement processes.

It will also adapt to the parallel system/hardware/software development process and the multiple software release cycle used within the automotive industry. The software release cycle can be summarized as: "Release-Change Request(or Problem Resolution)-Update Release" in a semi-continual cycle (multiple repetitions for one product, until replaced by a new product). While this type of multiple release cycle is also used for software in other industries, it is more complicated in the automotive industry due to the parallel hardware ECU development cycles. In addition, it is possible that during the prototype development stage, there are competing suppliers providing both hardware and software. This requires extra emphasis on processes covering software release and configuration management.

Due to the multiple release strategy adopted when developing software for automotive Electronic Control Units, Change Request Management is an important process to manage these changes, and configuration management becomes important in tracking the many versions of software released both during development and in the units that are included in the delivered automobiles.

6.5   Capability Maturity Model Integration and FAA-iCMM

The Software Engineering Institute's latest versions of CMM® are the Capability Maturity Model Integration (CMMI® v1.1) models. There are two types of CMMI® models, the continuous model and the staged model. The staged model is based upon the staged representation first formulated for the SW CMM®; in brief, it requires processes to be implemented in a specific order to attain a maturity level. This approach is not fully compatible with the requirements of ISO/IEC 15504 and will not be described further here. For more detail on this model, see the Software Engineering Institute CMMI® Handbook [33] or the CMMI® website [34].

The CMMI® provides both a Process Reference Model and Process Assessment Model. Because the Process Assessment Model encapsulates the reference model, it is described here, rather than in chapter 5.

The CMMI® - continuous model has been designed to be conformant to ISO/IEC 15504[34]. There are several variants of this model available, each presenting a different scope of activity. The most comprehensive model is briefly summarized here. This is the CMMI® for Systems Engineering, Software Engineering, and Integrated Product and Process Development [35].

The model consists of the following process areas.

- Process Management
  - Organizational Process Focus (B)
  - Organizational Process Definition (B)
  - Organizational Training (B)
  - Organizational Process Performance (A)
  - Organizational Innovation and Deployment (A)
- Project Management
  - Project Planning (B)
  - Project Monitoring and Control (B)
  - Supplier Agreement Management (B)
  - Integrated Project Management for IPPD (Integrated Product and Process Development) (A)
  - Risk Management (A)

---

[34] It should be noted that work is still ongoing to ensure a fully conformant model is achieved.

- Integrated Training (A)
- Integrated Supplier Management (A)
- Quantitative Project Management (A)
- Engineering
  - Requirements Management (B)
  - Requirements Development (B)
  - Technical Solution (B)
  - Product Integration (B)
  - Verification (B)
  - Validation (B)
- Support
  - Configuration Management (B)
  - Process and Product Quality Assurance (B)
  - Measurement and Analysis (B)
  - Decision Analysis and Resolution (A)
  - Organizational Environment for Integration (A)
  - Causal Analysis and Resolution (A)

The CMMI® continuous model groups processes into process areas to highlight the specific interactions, but the process areas also often interact and have an effect on one another regardless of their defined group.

The engineering process areas are written in a general engineering terminology so any technical discipline involved in the product development process (e.g., software engineering, mechanical engineering) can use them for process improvement. At the same time, the CMMI® is oriented towards organizations that develop systems and software, and not to organizations that do not do this.

The CMMI® also differentiates between basic process areas (shown as (B) in the process areas list) and advanced process areas (for example, Organizational Process Performance and Organizational Innovation and Deployment are advanced process areas – shown as (A) in the process areas list). It therefore guides users to consider implementing the basic process areas before the advanced process areas to have a good foundation to build upon – especially when attempting to achieve higher Capability Levels.

The layout of CMMI® provides for each process area:

- Purpose

- Introductory notes
- Related Process Areas
- Specific Goals
- Generic Goals
- Practice to Goals relationship
- Specific Practices by Goal
- Generic Practices by Goal

For the purpose of a conformant Process Reference Model, the purpose, specific and generic goals descriptions are sufficient. The goals provide the equivalent description to process outcomes. For example, the Process and Product Quality Assurance process area consists of the following description [36]:

Purpose
The purpose of Process and Product Quality Assurance is to provide staff and management with objective insight into processes and associated work products.
Specific Goals
SG 1 Objectively Evaluate Processes and Work Products [PA145.IG101]
Adherence of the performed process and associated work products and services to applicable process descriptions, standards, and procedures is objectively evaluated.
SG 2 Provide Objective Insight [PA145.IG102]
Noncompliance issues are objectively tracked and communicated, and resolution is ensured.
Generic Goals
GG 1 Achieve Specific Goals [CL102.GL101]
The process supports and enables achievement of the specific goals of the process area by transforming identifiable input work products to produce identifiable output work products.
GG 2 Institutionalize a Managed Process [CL103.GL101]
The process is institutionalized as a managed process.
GG 3 Institutionalize a Defined Process [CL104.GL101]
The process is institutionalized as a defined process.
GG 4 Institutionalize a Quantitatively Managed Process [CL105.GL101]
The process is institutionalized as a quantitatively managed process.
GG 5 Institutionalize an Optimising Process [CL106.GL101]
The process is institutionalized as an optimising process.
CMMI-SE/SW/IPPD, V1.1

Example CMMI process area description.

The model presented must be adapted and tailored by the organization to their business needs. The CMMI® provides guidance for users to transition from the older CMM® standards and EIA/IS 731 [37] to CMMI® with the aim to allow assessments to achieve ISO/IEC 15504 conformance.

6.6 US FAA iCMM

This section is based upon information kindly provided by Linda Ibrahim, the US FAA Chief Engineer for Process Improvement. I have edited the input to place it within the context of the book and added some comments comparing the FAA iCMM® to other process assessment standards, especially ISO/IEC 15504.

The United States Federal Aviation Administration (US FAA) has created its own model called the FAA iCMM® [38].

The US Federal Aviation Administration commenced process improvement using various forms of the SEI Capability Maturity Model. In 1997, they integrated three of the capability maturity models (CMMs): the software, systems engineering, and software acquisition CMMs. This became the integrated CMM (or FAA iCMM® v1.0 for short) and was published and deployed in 1997.

The US FAA soon recognized that integrating CMMs alone did not provide sufficient process improvement guidance for broad and complex enterprises like the FAA and others who have adopted the FAA iCMM. Thus, the US FAA created version 2.0 in 2001. This integrated several important improvement approaches including ISO/IEC TR 15504 (SPICE), ISO 9001:2000, ISO/IEC CD 15288, ISO/IEC 12207, Electronics Industries Alliance (EIA) 731, Malcolm Baldrige National Quality Award, and Capability Maturity Model Integration. By integrating beyond CMMs, and incorporating international standards and perspectives, the iCMM now contains cohesive guidance for improving business and technical processes ranging from enterprise management to operation and disposal.

The FAA iCMM continues to evolve, with work currently underway to include international safety and security standards as part of the framework. There are detailed mapping tables indicate how the practices and principles from source standards and documents are integrated into the FAA iCMM.

Any organization pursuing process improvement examines its business objectives and the processes it performs to accomplish those objectives. The organization selectively compares its practices to those in the relevant parts of the FAA iCMM to identify areas where improvements might be pursued.

The model provides process improvement coverage for internal and external organizations to the FAA. Its primary focus was initially internal and governmental organizations, but it is gaining wider acceptance. It covers high-level (for example strategic direction) and low-level (for example task management) issues. It can be used for process assessment of external organizations in combination with an FAA Assessment method [39].

General Structure of the FAA iCMM.

The practices in the iCMM are structured into 2 parts: the process dimension, with practices specific to performing a selection of processes, and the capability dimension. The FAA deliberately followed the recommended guidance for process assessment models in ISO/IEC 15504. The Base Practices are specific to each process, while the Generic Practices are used to improve the way any process is performed. Within the process dimension, practices are grouped into Process Areas, and within the capability dimension, practices are grouped into Capability Levels. Both Process Areas and Capability Levels contain goals expressing what should be achieved if their associated practices are performed. Some practices in the process dimension provide additional detail regarding practices in the capability dimension. This structure is illustrated below.

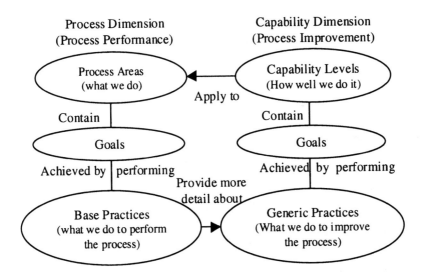

FAA iCMM Dimensions, Goals and Practices

The Capability Dimension.

There are 6 capability levels in the capability dimension of the iCMM. Any of the processes in the process dimension may be being performed at any of these capability levels.

FAA-iCMM Capability Dimension

FAA-iCMM Capability Dimension: Capability Levels, Goals, and Generic Practices
CAPABILITY LEVEL 0: INCOMPLETE
One or more of the goals of the process area are not achieved. (No goal or generic practices at this level)
CAPABILITY LEVEL 1: PERFORMED
Goal: The process achieves the goals of the process area.
Generic Practices
1.1 Identify Work Scope
1.2 Perform the Process
CAPABILITY LEVEL 2: MANAGED: PLANNED AND TRACKED
Goal: The process is institutionalized as a managed (planned and tracked) process.
Generic Practices
2.1 Establish Organizational Policy
2.2 Document the Process
2.3 Plan the Process
2.4 Provide Adequate Resources
2.5 Assign Responsibility
2.6 Ensure Skill and Knowledge
2.7 Establish Work Product Requirements
2.8 Consistently Use and Manage the Process
2.9 Manage Work Products
2.10 Objectively Assess Process Compliance.
2.11 Objectively Verify Work Products
2.12 Measure Performance
2.13 Review Performance with Higher-level Management
2.14 Take Corrective Action
2.15 Coordinate with Stakeholders
CAPABILITY LEVEL 3: DEFINED
Goal: The process is institutionalized as a defined process.
Generic Practices
3.1 Standardize the Process
3.2 Establish and Use a Defined Process
3.3 Improve Processes
CAPABILITY LEVEL 4: QUANTITATIVELY MANAGED
Goal: The process is institutionalized as a quantitatively managed process.
Generic Practice:
4.1 Stabilize Process Performance
CAPABILITY LEVEL 5: OPTIMIZING
Goal: The process is institutionalized as an optimizing process
Generic Practice:
5.1 Pursue Process Optimization

The capability dimension of the FAA iCMM uses a similar concept of continuous capability levels, but a different definition of each capability level when compared to ISO/IEC 15504.

The major differences include:

- At Capability Level 2, the following generic practices are specifically addressed:
  - Establish Organizational Policy
  - Document the Process
  - Ensure Skill and Knowledge
  - Review Performance with Higher-level Management
  - Coordinate with Stakeholders
  - Measure Performance
- At Capability Level 2, the following generic practices are further evolved than in ISO/IEC 15504:
  - Consistently Use and Manage the Process
  - Objectively Assess Process Compliance
  - Take Corrective Action
- At Capability Level 3, the following generic practice is addressed:
  - Improve Processes

Note: These differences in the practices for the capability levels makes it difficult to translate or compare the results of the FAA iCMM capability levels directly to ISO/IEC 15504 capability levels (and process attributes).

The Process Dimension.

- There are 23 process areas in the process dimension of the iCMM. Some of these process areas pertain to **Management** activities, some to **Life Cycle** activities, and some to **Support** activities.

Management

PA 00   Integrated Enterprise Management (ML3)

PA 11   Project Management (ML2)

PA 12   Supplier Agreement Management (ML2)

PA 13   Risk Management (ML3)

PA 14   Integrated Teaming (ML3)

Life Cycle

PA 01   Needs (ML3)

PA 02   Requirements (ML2)

PA 03   Design (ML3)

PA 06   Design Implementation (ML3)

PA 07   Integration (ML3)

PA 08   Evaluation (ML2)

PA 09   Deployment, Transition, and Disposal (ML2)

PA 10   Operation and Support (Not staged)

Support

PA 04   Alternatives Analysis (ML3)

PA 05   Outsourcing (ML2)

PA 15   Quality Assurance & Management (ML2)

PA 16   Configuration Management (ML2)

PA 17   Information Management (Not staged)

PA 18   Measurement and Analysis (ML2)

PA 20   Process Definition (ML3)

PA 21   Process Improvement (ML3)

PA 22   Training (ML3)

PA 23   Innovation (ML5)

Most of these process areas are assigned a maturity level designation (ML) providing guidance on which process areas might be improved together, or next. These designations also offer iCMM maturity levels for the purpose of benchmarking with organizations that measure their accomplishments using the maturity levels of a variety of CMMs that are sources to the iCMM. Since the iCMM integrates many standards and models besides CMMs however, some process areas that are not in other CMMs are designated as "not staged" for benchmarking purposes.

Maturity Level Definitions:

- Maturity Level 2: To achieve maturity level 2, nine process areas staged at maturity level 2 must have satisfied capability levels 1 and 2 (or be not applicable) according to an iCMM appraisal. See the process table below for the ML2 processes.

- Maturity Level 3: To achieve maturity level 3, 20 process areas staged at maturity level 2 and maturity level 3 must have satisfied capability levels 1, 2, and 3 (or be not applicable) according to an iCMM appraisal. See the process table below for the additional ML3 processes.
- Maturity Level 4: To achieve maturity level 4, 20 process areas staged at maturity level 2 and maturity level 3 must have satisfied capability levels 1, 2, and 3 (or be not applicable) and selected process areas additionally must have satisfied capability level 4 according to an iCMM appraisal.
- Maturity Level 5: To achieve maturity level 5, 21 process areas staged at maturity levels 2, 3, and 5 must have satisfied capability levels 1, 2, and 3 (or be not applicable) and selected process areas additionally must have satisfied capability levels 4 and 5 according to an iCMM appraisal.

Sample Process Description.

PA 00 INTEGRATED ENTERPRISE MANAGEMENT
Goals
1. Vision, mission, values, performance goals and objectives are established, maintained, and communicated to all employees. 2. Strategies are developed and projects are launched that visibly support goal achievement. 3. Projects are continued, changed, or terminated based on performance, within the capability of the organization, and with acceptable risk and potential benefit to the organization.
Practices
BP 00.01   Establish and Maintain Strategic Vision BP 00.02   Align to Achieve the Vision BP 00.03. Establish and Maintain Strategy BP 00.04. Develop and Deploy Action Plans BP 00.05. Review Performance BP 00.06. Act on Results of Review BP 00.07. Fulfill Public Responsibility

Note: One of the positive features of the FAA iCMM is that it contains processes at organization executive levels (for example, Integrated Enterprise Management as shown above), as well as processes like Outsourcing and Deployment, Transition, and Disposal. It does not assume that the enterprise is involved in software development.

Generic Attributes

The iCMM contains a feature used to measure the results of process performance, independently of capability level. There are 2 generic attributes defined:

- Usefulness: the extent to which work products or services provide the needed benefits in actual use
- Cost effectiveness: the extent to which the benefits received are worth the resources invested.

Note: The generic attributes are business related, rather than process assessment related and are a worthwhile feature of the model. This approach aligns well with the improvement focus.

For more information on how a business related improvement focus, provides tangible business benefits, readers are advised to read the related Practical Guide book, especially the improvement business case and the Team Based Business Design Improvement sections.

Some Advantages/Disadvantages.

The FAA iCMM provides an integration of various International and National standards into one model.

The model is being extended to cover specific areas of application. For example, it provides an 'application area' for safety and security related applications.

The model follows the ISO/IEC principle of allowing an organization to select only the processes relevant to its business needs. Thus an organization can decide to implement one process (or at least one process for a specific business area).

The model covers executive level management.

The model adopts the 6 capability level continuous measurement framework approach of ISO/IEC 15504.

However, the FAA iCMM uses a different capability dimension to ISO/IEC 15504, making assessments incompatible with the measurement framework of the International standard. This is probably the most serious disadvantage, if compatibility with the ISO/IEC 15504 measurement framework is critical.

The FAA iCMM and the SEI CMMI capability levels and maturity levels are also not completely compatible.

The FAA iCMM has integrated the Malcolm Baldrige National Quality Award criteria, which makes it useful for US institutions. However this also makes it somewhat incompatible with the European Quality Award criteria and the Japanese Quality Award criteria. This creates differences on a cultural basis, that should be recognised and handled.

There is currently no guidance on the desired capability levels for specific applications (for example, safety critical application).

6.7   Use of assessment indicators in rating processes

An assessment indicator is defined in ISO/IEC 15504 as an "objective attribute or characteristic of a practice or work product that supports the judgment of the performance or capability of an implemented process". The assessment indicators must provide a greater level of detail for each process attribute in order for the assessor to be able to rate that process attribute.

The Process Assessment Model provides sets of the process attributes and sets of their characteristics for each process in the model (Process Dimension). These are, in brief, the specific process purpose, outcomes and process attributes for each process (base practices) and the work products with descriptions of their content.

CAUTION: The process categories in a Process Assessment Model represent convenient groupings of processes for assessment purposes, but may not represent the actual process implementation of a business.

The name, format and allocation of the characteristics of the work products for each process are not mandatory. It is up to the assessor and the organizational unit coordinator to relate the actual work products produced in their organization to the associated work products for each process. Assessors need to convert/translate the process dimension to the actual business processes during assessment preparation.

The model also provides a set of indicators of each management practice consisting of extended descriptions containing the indicator description, the practice performance characteristics, the resource and infrastructure characteristics and the associated processes (Capability Dimension).

The assessment indicators in the model give examples of types of evidence that an assessor might obtain, or observe, in the course of an assessment. The assessors must map the actual objective evidence onto the set of assessment indicators to enable them to correlate the implemented process with the processes that are defined in the Process Assessment Model. The assessment indicators are guidance to the assessors to collect the objective evidence needed to support their judgment of capability; they are not a mandatory set of check-lists/products.

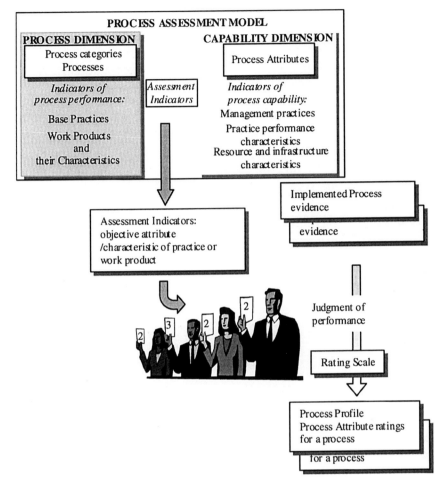

Use of Assessment Indicators and output of process profiles.

The output from a process assessment is a set of process profiles, one for each instance of each process within the scope of the assessment (it is possible to assess several instances of a particular process in one assessment, for example the assessors may assess the project management process in several projects).

Each process profile consists of a set of the process attribute ratings for an assessed process. Each process attribute rating represents a judgment by the assessor of the extent to which the attribute is achieved, and comprises the set of assessment indicators fulfilled by the accumulated evidence. This evidence may come either from the examination of work products of the processes assessed, or from statements made by the performers and managers of the processes.

The practical guide provides more detailed guidance upon the use of assessment indicators and the rating of processes in an assessment.

7 PROCESS LIFECYCLE MODELS

The Process Reference Models described earlier are not detailed enough to guide implementation of processes.

In this chapter, I look at some process lifecycle models used within industry that provide the extra level of detail and guidance needed to implement software development. The description in this chapter is only a starting point for readers to understand the basic characteristics of the 3 process lifecycle models written about. Readers should refer to the texts in the bibliography if they wish to know more or wish to implement any of these models.

In addition, this chapter briefly looks at how each process model relates to ISO/IEC 15504.

This chapter will be of interest to anyone designing processes, including process owners, process experts, and developers of process models.

7.1 German V-Model

Overview of V-Model

The V-Model is an internationally recognized standard [40] for development of IT systems for the German Federal government[35]. It comprises:

- the Lifecycle Process Model,
- the Allocation of Methods, and
- the Functional Tool Requirements.

The V-Model (1997) [41] specifies what steps are to be taken and which methods are to be applied for the development tasks and which functional characteristics the tools to be used must have. The V-Model realizes some of the processes in ISO/IEC 12207 in a specific and detailed manner [42].

The model is structured into three parts:

- Part 1: Regulations. This part contains binding regulations concerning work steps to be performed (activities) and results (products). This is the document set of interest for process assessment.

---

[35] The standard is non-proprietary and not copy-protected, which means that the V-model can be copied as often as is wished for one's own use without infringing license regulations.

- Part 2: Supplements with regard to Authorities. This part covers civilian and military fields with instructions to apply the Lifecycle Process Model in the respective field.

- Part 3: Collection of Manuals. This part contains a set of manuals dealing with special topics, such as IT security or use of object-oriented languages.

The V-Model was developed with several applications in mind, including as the basis for contracts, for instruction and for communication. The provisions of the V-model are independent of the form of organization. They are restricted to the technical development process. The V-model was developed as a standard in public administration and is used in the private sector in German banks, insurance companies, car manufacturers, manufacturing industry and energy producers.

The standard has the following three levels:

1. Procedure. "*What* has to be done?" This level establishes what activities are to be performed in the course of system development, which results are to be produced in this process and what contents these results must have.

2. Methods. "*How* is something to be done?" Here it is determined what methods are to be used to perform the activities established in the first level and what means of presentation are to be used in the results.

3. Tool Requirements. "*What* is to be used to do something?" At this level, what is established is what functional characteristics must the tools have which are to be used during system development.

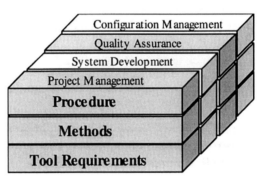

V-Model levels and areas of activity.

At all levels, the regulations are structured according to the areas of activity into sub-models that consist of Project Management, System Development, Quality Management and Configuration Management.

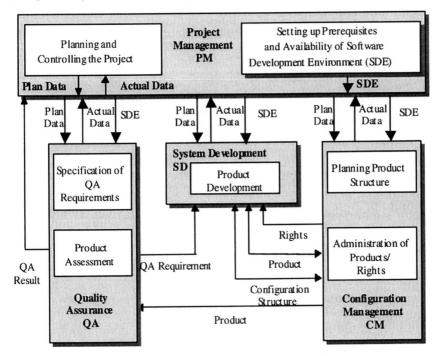

Coordination of V-Model activity areas (sub-models).

For each activity area, the V-Model provides a sub-model, activities and sub-activities, products and roles of personnel. When we look at the System Development sub-model, it comprises the following activities:

- System Requirements Analysis (SD 1): Description of the requirements for the system under development and its technical and organizational environment; realization of a threat and risk analysis; drafting a user-level model for functions, data and objects.
- System Design (SD 2): Segmentation of the system into segments and software and hardware units.
- Software/Hardware Requirements Analysis (SD 3): The technical requirements made for the software and, possibly, for the hardware units are detailed. From here onwards, the further progress is split into software development and hardware development.

- The software development of a software unit is performed in four steps/major activities (SD 4-SW to SD 7-SW).
- Similarly, the hardware development of a hardware unit is performed in four steps (SD 4-HW to SD 7-HW).
- System-Integration (SD 8): Integration of the various software and hardware units into a segment and integration of the segments (if they exist) into the system.
- Transition to Utilization (SD 9): Description of all activities that are necessary to install a finished system at the appointed point of operation and to put this into operation.

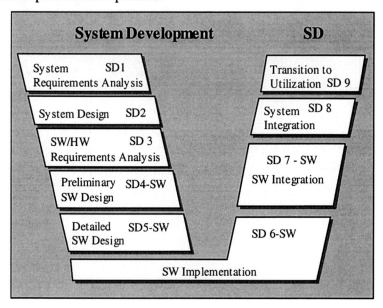

V-Model System Development Activities overview.

V-Model mapping to ISO/IEC 12207 and ISO/IEC 15504

The V-Model contains a comprehensive mapping shows conformance of the V-Model to parts of ISO/IEC 12207[36] (which is a conformant Process Reference Model for ISO/IEC 15504[37]. The V-Model mapping can be

---

[36] Please note that the mapping is against ISO/IEC 12207 and not against Amendment 1 of ISO/IEC 12207

[37] The mapping tables are taken directly from the section: The V-Model in the ISO and AQAP environment, of General Directive No. 250 "Lifecycle Process Model".

directly used with a process mapping of ISO/IEC 12207 to ISO/IEC 15504 for process assessments.

When performing an assessment, the assessors should work with the participating process personnel from the organizational unit to ensure that the process mapping is checked during the planning. When the V-Model is implemented consistently in an organization, the process mapping can be performed once for the entire organization and then used as needed in each process assessment.

Primary Lifecycle Processes

In the V-Model, the ISO/IEC 12207 Acquisition and Supply processes are partially covered by the V-model. The ISO/IEC 12207 Operation process has not been covered (although there is a limited coverage described).

The ISO/IEC 12207 development process can be completely realized with the V-model. The development process includes sub-model System Development (SD) (with the exception of the hardware) and parts of sub-model Quality Assurance (QA) of the V-model.

The ISO/IEC 12207 Maintenance process is covered in the V-model only with regard to general software maintenance and modification activities. The V-model does not cover the aspect "Software Retirement".

The following excerpt for the development process mapping from the V-Model illustrates the depth of detail that the model provides.

Allocation of Primary Lifecycle Engineering Processes to V-Model (sample)

Development Process		
Activities acc. to ISO/IEC 12207	Activities according to V-Model	
Process Implementation	PM1.3	"Generation of Project-Specific V-Model"
	PM1.4	"Toolset Management"
	PM 4	"Detailed Planning"
System Requirements Analysis	SD 1.2	"Description of Application System"
	SD 1.5	"User-Level System Structure"
	QA 4	"Product Assessment"
System Architectural Design	SD 2.1	"Technical System Design"
	SE 2.4	"Allocation of User Requirements"
	SD 2.5	"Interface Description"
	QA 4	"Product Assessment"
Software Requirements Analysis	SD 3	"SW/HW Requirements Analysis"
	QA 4	"Product Assessment"
	PM 6	"Phase Review"
Software Architectural Design	SD 4.1-SW	"SW Architecture Design"
	SD 4.2-SW	"Design of Internal and External SW Interfaces"
	PM 6	"Phase Review"
Software Detailed Design	SD 5.1– SW	"Description of SW Component/Module/Database"
	QA 1.2	"Generation of Assessment Plan"
	PM 6	"Phase Review"
Software Coding and Testing	SD 6.1-SW	"Coding of SW Modules"
	SD 6.2-SW	"Realization of Database"
	QA 2.3	"Definition of Test Cases"
	QA 4	"Product Assessment"
Software Integration	SD 4.3-SW	"Specification of SW Integration"
	SD 7-SW	"SW Integration"
	PM 6	"Phase Review"
Software Qualification Testing	QA 4	"Product Assessment"
	CM 2	"Product and Configuration Management"
	PM 6	"Phase Review"
System Integration	SD 8	"System Integration"
	QA 2.3	"Definition of Test Cases"
	QA 4	"Product Assessment"
System Qualification Testing	QA 4	"Product Assessment"
	PM 6	"Phase Review"
Software Installation	SD 9	"Transition to Utilization"
	PM 6	"Phase Review"

Each of the activities in the table is linked to an extensive description.

Supporting Processes

The V-model does not include any separate documentation activity. *Documentation* is not contained in the realization of activities. Therefore, the allocation of V-model activities is not possible, even though the documentation process in the entire V-model is naturally completely realized.

Apart from the configuration management and the change management of the V-model, the *Configuration Management* process of the ISO/IEC 12207 standard also handles "Configuration Evaluation". There is no similar description in the V-model.

According to ISO/IEC 12207, the *Joint Review, Audit, Verification* and *Validation* are handled as techniques of the process *Quality Assurance*. However, since the quality assurance process has only been roughly described, an explicit placement of these techniques is missing. Though the V-model covers all of the listed QA aspects, a 1:1 allocation between V-model and ISO/IEC 12207 is not always possible, based on the general information given in the ISO/IEC 12207 and the mixture between activities (tasks) and methods (techniques).

The V-model always distinguishes between the activities process assessment, product assessment and phase review, and, on pages with method allocations, between audit, review (technical review and phase review), and testing, whereby the reason for applying a certain method depends on the objects tested (software, product, activity, project development).

Organizational Processes

The two processes *Management* and *Infrastructure* are covered by the V-model, sub-model PM. The V-model does not contain an actual *Improvement* process; however, this can be achieved by means of process assessments and the resource project history, where results can be derived for the improvement process.

The V-model includes training and introduction information for staff members and personnel for the *Training* process.

In summary, the V-Model mapping is a good example of detailed process mapping against the ISO/IEC 12207 Process Reference Model and provides a reliable basis for process assessment.

7.2   Rational Unified Process

The Rational Unified Process ® (RUP) [43] is a software engineering process architecture that is iterative, use case driven and system architecture centric. It is founded on a process architecture that provides commonality across a family of processes. It is a configurable process, and should be tailored to various size project teams.

It provides an approach that assigns tasks and responsibilities in a (project) team, comprising requirements management, design, test, project management and configuration management. It provides project teams with access to a knowledge base with guidelines, templates and tools for all critical development activities. When the team accesses the same knowledge base, which uses common language and processes, it helps to create a team view. This tool support for the knowledge base is an important productivity and management component of the Rational Unified Process.

It focuses on addressing high-risk areas early in system/software development including defining a system architecture. The overall approach uses iterative requirements analysis/definition and development. To manage the changes implied in an iterative approach without losing control, it encompasses change control processes (supported by tools) to manage change.

The activities in the Rational Unified Process develop and maintain models (using Unified Modelling Language nomenclature), which are semantically rich representations of the software system under development.

The Rational Unified Process is supported by several products, which automate parts of the process. They are used to create and maintain the various artefacts—models in particular—of the software engineering process including the visual models, programming, testing, etc.

The Rational Unified Process encompasses the overall process attributes required for a Capability Level 3 process – it specifies both a standard and a defined process. I will briefly describe later what is required in implementation to achieve this Capability Level.

The main practices and process lifecycles are briefly described here, but readers should consult the numerous texts and books from IBM® and Rational®.

Practices:

- Develop software iteratively, thereby increasing understanding of the problem through successive refinements. Incrementally grow an effective solution over multiple iterations. The aim of this iterative refinement and incremental development is to reduce risk through frequent, executable releases that enable continuous end user involvement and feedback. Implementations may have release cycles as short as 2 weeks, although larger projects often use a 2, 3 or 6 month release cycles. The number of iterations may be between 3 and 9. These iterations and release cycle times are not fixed by the process itself, but rather by the customer requirements and the project team implementation.

- Manage requirements through a requirements elicitation process with the customer. Manage and document the required functionality and constraints. Track and document tradeoffs and decisions made during the elicitation process. The customer's business requirements are captured in use cases and scenarios (Business Modelling).

- Use component-based architectures with early development and base lining of a robust executable architecture (this is often incomplete and also iteratively developed). Use iterative and incremental component-based software development.

- Visually model the software to capture the structure and behaviour of architectures and components. If visual abstraction is correctly performed, it helps show how the elements of the system fit together. When the building blocks are consistent with the code, it helps to maintain consistency between a design and its implementation. Visual modelling can also help promote unambiguous communication when described using the industry standard Unified Modelling Language (UML).

- Verify software quality with respect to the requirements based on reliability, functionality, application performance and system performance.

- The process assists in the planning, design, implementation, execution, and evaluation of test types, using objective measurements and criteria.

- Control changes to software including the ability to make certain that each change is acceptable, and being able to track changes (including back-tracking).

- The Rational Unified Process provides individual and team workspace by describing how to automate integration and build management.

The Rational Unified Process can be described in two dimensions, or along two axes:

- The horizontal axis represents time and shows the dynamic aspect of the process as it is enacted, and it is expressed in terms of cycles, phases, iterations, and milestones.
- The vertical axis represents the static aspect of the process: how it is described in terms of activities, artefacts, workers and workflows. This dimension is the corollary of the process areas/processes in the process dimension of ISO/IEC 15504, but not directly equivalent.

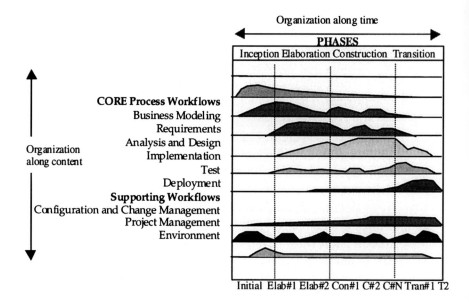

The Rational Unified Process dimensions.

The static aspects of the process in terms of activities and workflows are further elaborated with the Rational Unified Process. For example, the Analysis and Design Process consists of 6 sets of activities as shown in the diagram.

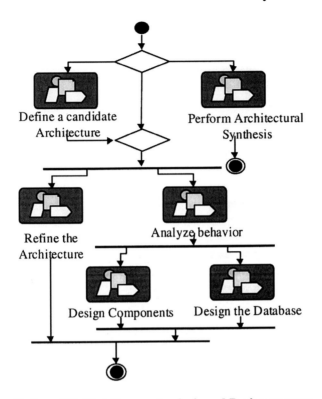

Rational Unified Process Analysis and Design process.

The Rational Unified Process uses a process composition that can be mapped to a Process Reference Model such as ISO/IEC 12207 and subsequently to a Process Assessment Model such as ISO/IEC TR 15504-5.

The following information shows the Rational Unified Process mapping to the Process Assessment Model in ISO/IEC TR 15504. The processes were mapped to the equivalent CUS, ENG, SUP, MAN and ORG processes, and then compared with the capability dimension. The results come from a comprehensive report prepared by Rational Software [44]. The report provides an analysis of the processes to ISO/IEC 15504 process assessment processes and work products, and an analysis of the capability dimension process attributes to management practice (MP) detail level.

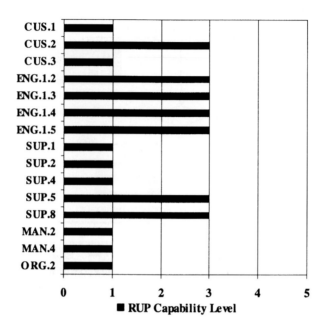

Rational Unified Process to ISO/IEC 15504 processes and Capability Levels.

For example, the Analysis and Design process shown earlier is mapped to ENG.1.2, ENG.1.3, ENG.1.4 and ENG.1.5.

The process mapping and capability determination can generally be considered conservative when assessing the potential of a process to achieve a particular Capability Level. In the above figure, the Capability Level rating is shown when the level is fully or largely achieved. The report notes that these results are the *potential* Capability Levels if all the processes are implemented as specified, so tailored implementations may have a different process capability.

At Capability Level 1, the Rational Unified Process is prescriptive enough to be able to find clear process mappings to most process assessment base practices and work products.

The second aspect is how some processes (when properly implemented) would lead to a rating at Capability Level 3.

The Rational Unified Process provides a lot of process implementation guidance. If this guidance were followed, then the process would reach Capability Level 3. For example, if the guidance is followed for the Analysis and Design process, then an instance of *any* process component (ENG.1.2,

ENG.1.3, etc), which was individually assessed as fully achieving Level 1, could also achieve Level 3. Since the Rational Unified Process is meant to be tailored for a project, Rational could not explicitly state that any process instance would reach Capability Level 3, but the potential exists for those processes shown (CUS.2, SUP.5, SUP.8 and ENG.1.2 to ENG.1.5) to reach Capability Level 3.

While the Rational Unified Process was assessed against the technical report version of the standard - ISO/IEC TR 15504, it is possible to update this in general for the published version of the standard as I have done here.

Looking at the ISO/IEC 15504:2003 version of the standard, the following can be deduced about process attributes that the Rational Unified Process does not completely specify and hence fulfil.

PA 3.1 Process definition attribute

b) the sequence and interaction of the standard process with other processes is determined;

c) the required competencies, roles, responsibilities and authorities for performing a defined process are identified as part of the standard process;

e) appropriate data are collected and evaluated to determine where continual improvement of the standard process can be made.

PA 3.2 Process deployment attribute

b) the required roles, responsibilities and authorities for performing the defined process are assigned and communicated;

c) the personnel performing the defined process are competent on the basis of appropriate education, training, skills and experience;

f) appropriate data are determined, collected and analysed as a basis for understanding the behaviour of, and to demonstrate the suitability and effectiveness of the defined process.

[ISO/IEC 15504-2 5.4]

PA 3.1 b) is fulfilled for the processes described within the Rational Unified Process, but (naturally) not to the other processes that an organization must employ. For example, an interaction with a separate Human Resource Management process is needed to meet the requirements of PA 3.1 c) and PA 3.2 b) and PA 3.2 c).

For PA 3.2 f) the Rational Unified Process does not explicitly specify a sizing or estimation method as a source of appropriate data, but does recommend use of methods such as COCOMO 2.0. Finally the Rational Unified Process specifies the iterations to refine the defined process used for a particular project, but does not mandate the feedback of this to the standard process as required by PA 3.1 e). This means that the rating would remain Largely achieved at Capability Level 3, mainly due to the intentional (limited) coverage of the Rational Unified Process, focused on software engineering.

It is therefore recommended that these issues are addressed in implementation of the Rational Unified Process, which would then potentially fully achieve Capability Level 3, and as a minimum partially achieve Capability Level 4 when COCOMO 2.0 is used (the main process attributes lacking at Capability Level 4 are for PA 4.2 Process Control).

In the next section, I describe eXtreme Programming. For readers interested in a comparison between RUP and XP from Rational's viewpoint, see the paper: A comparison of the IBM Rational Unified Process and eXtreme Programming [45].

7.3   Agile Development Methodologies

There are several well-known agile development methodologies in use. These include eXtreme Programming, Crystal methodologies (Crystal light, Crystal Orange and so on), and Dynamic System Development Methodology (DSDM).

Common aims of these methodologies are to reduce the amount of process specification, minimise process deployment and artefacts (work products). They do this in various ways, for example Alistair Cockburn describes several different weight Crystal methodologies depending mainly upon project size (workers and duration) and product type [46]. Each of these variants of Crystal is then tailored by the team at the start of the first iteration. Kent Beck describes a specific methodology for eXtreme Programming (see next section). Dynamic System Development Methodology specifies time box development tied to Joint Application Design (JAD) and Rapid Application Development (RAD). Timeboxing sets a specific (short) development iteration interval, at the end of which a running product is available.

All the agile methods emphasize people and communication with the aim to achieve a lighter weight process chain.

ISO/IEC 15504, due to its flexibility in specifying the process dimension can be effectively used with agile methodologies, when carefully adapted.

In one of my previous employers, where we ran several parallel projects, we originally used an incremental iterative development methodology created internally. This was specification intensive and could be considered a heavyweight methodology. Project teams were generally moderately large (40 to 140 people). Formal customer deliveries were 12 to 18 months apart. Both due to the strong (and highly effective) emphasis upon reuse of all developmental artefacts, and the type of systems produced, this methodology was very effective and successful in building large, high performance software systems.

Several of the processes (those we considered most important to successful projects) were rated at ISO/IEC TR 15504 capability level 4 and most were rated at capability level 3. The organizational culture was very quality focussed (evidenced firstly by high customer satisfaction, secondly by our ability to meet business goals, thirdly by our internal quality improvement programmes and finally by ISO certification and the high capability levels of our processes).

However we were asked to adopt a time to market driven approach by our of our customers. This necessitated adopting an agile methodology.

There was an overall project team, which comprised a customer requirements elicitation team, a project management and contracts team, between 3 and 5 software product development teams operating in parallel, and a deployment support team. After specifying a product in a user requirements workshop, the project and contract team made an offer and if accepted by the client, created a contract for that product. This triggered a development team with a fixed project duration of 6 months and two formal deliveries (with normally 4 to 6 complete internal development iterations). If the client wished to develop the product further, the contract could be extended or another contract created. The scope for the product development was prioritised and after the first 3-month delivery could be re-prioritised (or even another user requirements workshop could be run). As time to market was critical, scope was flexible, but as many prioritised requirements as possible were met within the contracted effort for the 6 months period. In some cases the client was co-located with the team, in other cases, the deployment team was handed the client from the elicitation team. Depending upon client location, requirements change was 3 monthly (externally located) or monthly (when co-located and running an internal monthly development iteration).

Each of the teams required an agile process. This included an agile development process, an agile client product elicitation process, agile contracting process, and agile deployment and support processes.

Due to our advanced quality culture and understanding of high capability processes, we were able to implement an agile process methodology (DSDM plus proprietary agile processes) in a very short time. Within one delivery cycle (3 months) the teams reviewed and improved the processes to a situation meeting our business goals, reaching our target capability and also ensuring process stability. (Process stability makes it easier to introduce new members to the project.) Naturally, it also met the client's needs in terms of time to market, but also in terms of client satisfaction with the products produced against the contracted (paid) effort.

We found that understanding ISO/IEC TR 15504 helped us to implement high capability processes in a short time. We did this using a techniques I invented called Business Process Mapping (see Practical Guide).

In order to assess the capability we had to then map our processes to the assessment model, as there was no exact correlation between ISO/IEC TR 15504-5 base practices, and the actual practices implemented in the project.

However, I highly recommend that the team creating the processes has both business experts and process capability (ISO/IEC 15504) experts, as well as practitioners. In the cited project, I acted as the process capability expert and was the mentor and coach to the teams.

It is important to carefully adapt process assessment to suit agile methodologies. It is possible to achieve this with ISO/IEC 15504 due to its flexibility in the process dimension, but it also requires deep understanding of the capability dimension.

I will describe this further for eXtreme Programming.

7.4   eXtreme Programming

eXtreme Programming (XP) is an agile development methodology that focuses on embracing change by planning, analysing and designing software in very small iterative increments [47].

The methodology allows for highly volatile requirements, through the planning games and short development intervals. It is highly suited to smaller development teams using short verbal communication paths within the team and with the customer (in other words, immediacy of communication and feedback). It results in minimal written documentation, simpler and smaller design and programming increments.

Kent Beck describes eXtreme Programming as a highly social activity (people oriented), rather than a documentation driven process in his book: eXtreme Programming Explained – Embrace Change [48]. This important philosophical distinction shapes many aspects of eXtreme Programming.

In the following figure, I highlight the 3 factors that contribute to organizational and project success, using the People-Process-Product model.

Compared to other well-established process lifecycles such as the V-Model and the Rational Unified Process, eXtreme Programming dramatically shrinks the role of processes (greatly simplifying but not totally eliminating them) and places much greater emphasis on the role of people in a team, hence highlighting team social and cultural aspects.

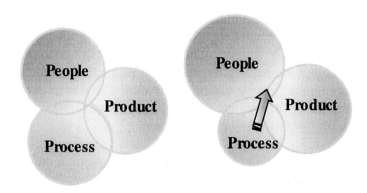

People-process-product comparison: process oriented and eXtreme Programming

In order to determine whether process assessment can add value in an eXtreme Programming environment, it is important to understand the basics of eXtreme Programming.

Kent Beck describes in his book values and principles that he believes are the foundation for success within this social activity context. The values need to be appealing to people so they readily adopt and share them, and yet still meet the organization's commercial (business) needs. The main shared values are:

✓ Communication,

✓ Simplicity,

✓ Feedback,

✓ Courage, and

✓ Respect.

Communication is a necessary value in order to make many of the principles and practices work effectively and efficiently. A shared value of communication leads to common understanding with the team.

Simplicity means it is better to make simple solutions today (and further simple changes in the future as needed) rather than design something more complicated, which may never be used. This is based upon two premises. First, that the cost of change does not increase exponentially over time (see the book for further explanation). Second, that changing requirements may invalidate many design decisions made earlier.

The shared value of feedback is concerned with immediate, concrete feedback as well as longer-term feedback to the team. Immediate, concrete feedback to developers is provided by performing the already written tests (Kent Beck describes this approach in his book: Test Driven Development [49]).

Courage is about taking the simple path (simple changes), but also deciding when code is bad and should be completely rewritten. When shared communication and feedback are effective, then courage combines with simplicity to provide the team with the ability to effectively make changes. Finally, the combination of these four values leads to a shared value of respect within the team.

(Readers wishing to know more about the importance and benefits of shared values within a social and cultural context should look at the chapter on Improvement and Culture in the Practical Guide.)

He then describes sets of principles derived from these values. They include:

- Rapid feedback.
- Assume simplicity.
- Incremental change.
- Embracing change.
- Quality work.
- Teach learning
- Small initial investment.
- Play to win.
- Concrete experiments.
- Open, honest communication
- Work with people's instincts, not against them.
- Accepted responsibility.
- Local adaptation.
- Travel light.
- Honest measurement

In Extreme Programming Explained and in his second book: Planning Extreme Programming [50], Kent Beck explains in his typically clear and simple manner (the 2nd value and principle) how these values and principles provide the primary foundation upon which the rest of XP is based.

When he describes aspects such as, courage, people's instincts, honest communication, travel light, and teach learning, he is describing both values and modes of behaviour he believes that the team must first accept and embrace in order to make eXtreme Programming succeed.

Readers are encouraged to read the book to understand the importance of the values and principles, so that the following 12 practices are placed within the team social context of shared values and principles.

Planning game. Customers decide the scope and timing of releases based on estimates provided by programmers. Programmers implement only the functionality demanded by the stories in this iteration.

Small releases. The system is put into production in a few months, before solving the whole problem. New releases are made often—anywhere from daily to monthly.

Metaphor. The shape of the system is defined by a metaphor or set of metaphors shared between the customer and programmers.

Simple design. At every moment, the design runs all the tests, communicates everything the programmers want to communicate, contains no duplicate code, and has the fewest possible classes and methods. This rule can be summarized as, "Say everything once and only once".

Tests. Programmers write unit tests minute by minute. These tests are collected and they must all run correctly. Customers write functional tests for the stories in an iteration. These tests should also all run, although practically speaking, sometimes a business decision must be made comparing the cost of shipping a known defect and the cost of delay.

Refactoring. The design of the system is evolved through transformations of the existing design that keep all the tests running.

Pair programming. All production code is written by two people at one screen/keyboard/mouse.

Continuous integration. New code is integrated with the current system after no more than a few hours. When integrating, the system is built from scratch and all tests must pass or the changes are discarded.

Collective ownership. Every programmer improves any code anywhere in the system at any time if they see the opportunity.

On-site customer. A customer sits with the team full-time.

40-hour weeks. No one can work a second consecutive week of overtime. Even isolated overtime used too frequently is a sign of deeper problems that must be addressed.

Open workspace. The team works in a large room with small cubicles around the periphery. Pair programmers work on computers set up in the centre.

Kent Beck emphasizes that the 12 practices should not be considered separately but as a complete set of practices. He emphasizes that the interactions between the practices are just as important as the practices themselves, and act as a self-reinforcing whole.

I add one important rule that Kent Beck emphasizes in his book.

Just rules. By being part of an eXtreme team, you sign up to follow the rules. But they're just the rules. The team can change the rules at any time as long as they agree on how they will assess the effects of the change.

Studying eXtreme Programming values and principles in practice

One study of a company using eXtreme Programming [51] has focused on the deeper cultural values and principles mentioned in eXtreme Programming and how they operate. The study derived and highlighted 5 themes that provide the basis for how the company uses eXtreme Programming. The paper emphasizes that an effective eXtreme Programming team needs to display social and cultural behaviour that achieves the purpose of these themes. The themes are:

Shared purpose, understanding and responsibility – decisions are communal and collective; programmers changed pairs frequently so that everyone paired with another programmer at least once during a iteration of the development, and also could overhear other pair's discussion and join in as interested; the oral tradition meant story cards were transient and were used until understood and then discarded/torn up and the metaphor was important to maintain clear communication. Note: the discarding of written evidence on the cards makes the process assessment more difficult but not impossible (both due to lack of work products and (nearly total) reliance on interviews as the real source of evidence).

Coding and Quality of code matters – coding was considered the most important activity that should not be interrupted, hence a designated 'exposed pair' handled any customer enquiries; test first was always implemented; tension arose when refactoring was not allowed within the story card estimated work; and code was not released after 5 pm due to poor quality experiences in the past.

Sustainability – the team cared about the quality of life, there were no heated discussions; retrospectives (after completion meetings) used a fun referee (toy barking dog) to create moderate behaviour in the speaker; there were regular, communal breaks; and the team used a toy 'moo' box to announce code release.

Harmonious Rhythm – the team atmosphere was calm, competent and confident; discussions and meetings (e.g. planning game, stand-up meetings) started when enough people present and ended informally when people left; the daily rhythm fit inside the 3 week iteration rhythm.

Fluidity – the team members were fluid in their work allocation and retention of common ownership; physical space allowed fluid movement; there were recognized boundaries of different roles and responsibilities but members changed roles regularly.

The theme of shared purpose, understanding and responsibility highlights the team approach. In ISO/IEC 15504 at Capability Level 2, it states that "responsibilities and authorities for performing the process are defined, assigned and communicated", in eXtreme Programming there is an assigned customer and tester, the other responsibilities and authorities are assigned to the team as a whole, and at any time different persons can assume different responsibilities and authorities. This is not incompatible with ISO/IEC 15504 but highlights the need for an assessor to carefully interpret the process attributes in ISO/IEC 15504.

Themes such as Harmonious Rhythm and Fluidity emphasize the difference between a process orientation to a team social and cultural orientation. A process assessor needs to recognize and understand how these social and cultural orientations change the way a process is implemented.

When Kent Beck states that the team members should only work a 40 hour week (sustainability), he emphasizes that if any overtime occurs for more than one week, then the project is in trouble and needs to go through new planning. This new planning in the planning game needs to look at scope, most often scope reduction and setting new priorities (shared responsibility).

All of the above values, principles and the way they are implemented (for example, as described themes) require careful understanding in order to 'translate' eXtreme Programming to a context where process assessment becomes viable and useful.

eXtreme Programming, CMM and ISO/IEC 15504

The Software Engineering Institute has looked at how eXtreme Programming can be related to the CMM® . Mark Paulk [52] has provided an overview mapping to the CMM® in an attempt to address the debate about whether they are compatible and can co-exist, and he concludes that compatibilities allow CMM® to be applied. One issue that still troubles people is that the staged version of CMM® requires specific processes for each maturity level and this creates conflicts between the two at level 2 and above.

In the following table, I 'map' the implementation of the eXtreme Programming practices to several processes defined within the ISO/IEC 15504 compatible process reference and process assessment models.

While a comprehensive process mapping is beyond the scope of this book, the following table shows some of the main relationships between eXtreme Programming practices and ISO/IEC 15504 process descriptions (e.g. from the ISO/IEC 12207 PRM). This is a broad interpretation of relationships, it is not an exact equivalence due to the interdependent nature of the 12 practices and the relationship to values and principles.

Overview between eXtreme Programming Practices and ISO/IEC 15504 Processes

EXtreme Programming Practice	Related ISO/IEC 15504 Process
Planning stories, onsite customer, Continuous integration	Requirements Elicitation Requirements Management
Planning game, stories, small releases, project velocity	(Software) Project Management
Pair programming	Review (peer review)
Collective Ownership of code, small releases and continuous integration	Configuration Management
Metaphor, Simple Design, Refactoring, Pair Programming.	Software Design
Testing (unit and functional)	Software Testing
Continuous Integration.	Software Integration and Test
Coding Standard[38]	Software Construction
40-hour Week.	No equivalent

Even though I show the relationships, there are challenges in more precisely mapping eXtreme Programming to ISO/IEC 15504, including:

---

[38] A coding standard is implied by collective ownership and pair programming, otherwise it is not possible for a programmer to improve code anytime he/she sees an opportunity. It does not imply that the coding standard is formally documented.

eXtreme Programming	ISO/IEC 15504
• technical work oriented	• broader management orientation
• oriented towards smaller projects with minimal documentation	• oriented towards larger projects with an emphasis on comprehensive documentation
• informal verbal communication means and information sharing	• formal means of communication and documentation
• few work products (code, tests) and fewer records of implementation	• many work products and records specified in the models
• larger project scaling open to debate [53] [54]	• small projects require extensive tailoring
• culture is personally and verbally oriented within a team	• culture is process oriented
• team driven selection of social interaction and practices	• management driven process oriented methodology (e.g. first a standard process (organizational), then a defined process for each team)
• discipline is exerted within the team by peer pressure	• discipline is (theoretically) organization management driven[39]

In Kent Beck's book, he has a diagram showing the relationship between the various practices that illustrates the inexact nature of the relationships in the above table. Pair Programming depends upon (and affects):

• Collective ownership.
• Coding standard.
• Continuous Integration.
• Metaphor.
• Refactoring.

---

[39] ISO/IEC 15504 specifies many organization and management processes, some of which support higher capability levels.

- Simple Design.
- Testing.
- 40-hour week.

Kent Beck emphasizes that the 'richness' in eXtreme Programming comes from the way these aspects interact (i.e. a holistic view). What he is saying is that the team implementing eXtreme Programming must adopt all the values, principles and practices (in that order) to make it work.

If software construction is assessed as a process, we need to look at aspects that are relatively easy to assess such as the coding standard, but also at how all the aspects together enhance communication (one of the main values and principles) to make pair programming effective and efficient as a software construction process. Therefore, the assessor needs to decide which of the base practices are relevant (and from which processes in a PAM).

The lack of records (particularly permanent records) in eXtreme Programming makes process assessment more difficult. As an example, difficulties include finding records that meet the process attributes at Capability Level 2 (for example planning records).

Since permanent record capture is not required in eXtreme Programming, an organization that wishes to do so needs to consider how to do this without imposing additional restraints or effort upon the team. There have been studies to look into how to automatically capture sufficient records to track a project history to aid a process assessment [55], showing it is possible to do this, without impacting greatly upon the project team.

One possible assistance to address the lack of management coverage in eXtreme Programming is the adoption of SCRUM™ to handle more of the management aspects. This is called xP@Scrum [56]. This provides a method that integrates well with eXtreme Programming – some aspects of Scrum are also used in eXtreme Programming, for example, the Sprint Planning Meeting in Scrum shares similarities with the Planning Game.

One study [57] of the use of xP@Scrum highlights how the combination covers management aspects. In this report on xP@Scrum, the organization is creating its own combination of eXtreme Programming, Scrum, CMM and ISO 9000 compliant approach. The organization has added a broader management and quality process framework to eXtreme Programming. This amalgamation of eXtreme Programming with ISO standards or organization management procedures is becoming more common as organization management wishes to ensure that management reporting and controls meet their requirements.

In situations where eXtreme Programming practices are adopted within a broader process approach, process assessment using ISO/IEC 15504 can provide value. For example, determining how well eXtreme Programming meets the customer and project goals using metrics or qualitative means is one area that can be assessed using the process attributes in an assessment.

Process assessors need to modify their approach when assessing an XP project. It is not sufficient to just collect data in an XP project against an existing Process Assessment Model.

For example the exemplar assessment model in ISO/IEC TR15504-5 has 40 Base Practices (BPs) in the software development [ENG] category (and another 7 for maintenance), and specifies a total of 76 input and 60 output work products (in reality many are repeated, so there are less actual work products). Since eXtreme Programming specifies 12 practices and many are very different to the base practices, the assessor needs to select the relevant eXtreme Programming practices and (minimal) work products and the relevant process attributes and base practices in a process assessment model.

The assessor needs to also understand how the team's culture, use of communication and informal information sharing complements/replaces some base practices and management practices. Currently there exists no detailed mapping of eXtreme Programming to any recognized process reference model or process assessment model. This means that it is more difficult to perform a conformant assessment.

An assessor therefore needs to understand the principles in eXtreme Programming, the 12 practices, and how the team applies then in their team culture in order to make a valid 'mapping' or translation between the implementation and a process assessment model, in order to perform an assessment.

To further complicate the assessor's task, at XP2003, Alan Francis (quoting Kent Beck) reported on the so-called XP-Maturity Model [58]. This consists of 3 levels:

- Level 0: follow all 12 practices all the time
- Level 1: modify the practices to suit the environment
- Level 2: it doesn't matter what you're doing as long as it works (see 'Just rules')

The 3 levels illustrate the basic cultural differences between eXtreme Programming and process-oriented methodologies, and the usual basis for process assessment. The suggested maturity levels are not aligned with ISO/IEC 15504 Capability Levels.

Level 2 in this XP Maturity Model is a prescription for extreme innovation and improvement, or extreme disaster – depending upon the level of discipline within the team (based upon collective ownership, respect, communication, courage, etc.).

The 'Just rules' rule is one of the more controversial aspects of eXtreme Programming. Opponents see it as an avenue for programmers to be hackers, proponents see it as an avenue to create strategic business advantage by optimising and further innovating the team approach and methods (which are ISO/IEC 15504 Capability Level 5 process attributes). Unfortunately, there is no one correct answer to this issue – a lot depends upon the discipline of the team – something that can worry management especially when first starting an eXtreme Programming project.

The range of implementations of eXtreme Programming is wide and some organizations are adopting only a few of the 12 practices within a process-oriented management approach. There have been other attempts to suggest alternate maturity models for eXtreme Programming, including one prepared by Poznan University in Poland [59]. This model looks at an order of implementation of the 12 practices.

Kent Beck expresses concern in his book about partial adoption, and states that this is no longer eXtreme Programming. The reason he provides is that the organisation first needs to adopt all the practices (and the values and principles) and successfully use them before it has sufficient understanding to change them. In other words: the 'No Rules' rule only applies once you have learnt how to apply eXtreme Programming properly and successfully. The reason is that the values and principles of eXtreme Programming must be the adopted basis for making decisions upon implementation of practices.

The cultural differences between process-oriented approaches and the social-cultural approach in eXtreme Programming means that process implementation may occur on a team-by-team basis, rather than an organization wide basis (a lot depends upon how much the experienced eXtreme Programming personnel including coaches move around between teams). Hence, process assessment has to recognize that team differences may be greater than in a more process-oriented approach.

Using process assessment to compare the way that eXtreme Programming is implemented in different project teams within one organization can therefore be a benefit to each team, as well as providing feedback to management.

Recommendations on the use of ISO/IEC 15504 process assessment for teams embracing eXtreme Programming and for assessors performing process assessments:

- Process assessment is more likely to be useful when eXtreme Programming is combined with some process-oriented methodology, rather than in a eXtreme Programming maturity level 2 team.

- Focus on a minimum set of software development processes (analysis and design, coding and testing), configuration management and customer requirements elicitation processes. When performing process mapping, use the most appropriate base practices to match the '12 practices', even if the base practices come from several different processes (in other words, don't restrict a process assessment to the base practices within a particular process but delete inapplicable practices and add applicable practices from other processes).

- Perform extensive (extreme) tailoring of the process attributes and indicators (base practices, generic practice indicators, generic work product indicators, generic resource indicators) to suit the team's project environment. In other words, it may be necessary to adapt the capability scale to suit the use of eXtreme Programming.

- Assessments should rely predominantly on interviews – using a team interview approach. (In any case, it is likely that the team will insist on this themselves).

- If an automated record-capturing product is implemented, use this as crosschecking evidence only when there is a dispute on how the method is implemented.

- Aim to use the assessment to provide value to the team and management. Comparative assessments between several teams in larger organizations can be useful.

- Try to capture the social and cultural aspects that indicate team cohesiveness and lead to team success, and/or note those aspects that are detrimental.

- Be open-minded and make the assessment enjoyable to all participants.

Given the communication and information sharing orientation inherent in the social activity approach of eXtreme Programming, there should be consideration given to creation of a specific process assessment model that suits this and other agile development methodologies.

In the meantime, the use of explicit mapping prior to an assessment is more likely to lead to repeatable assessments, than a reliance upon each assessor's judgment of what evidence to consider.

There is no doubt that eXtreme Programming is still a lively subject for debate within the software and process standards communities and will continue to be for some time.

In future editions of this book, subject to the interest of readers (so contact me), I am willing to investigate how other agile methodologies can benefit from process assessment.

8 ASSESSOR COMPETENCE AND BUILDING ASSESSMENT TEAMS

The chapter describes at the personal requirements for persons to become competent assessors. In addition, it looks at the registration and certification requirements for assessors. Finally, it briefly describes the requirements when assembling an assessment team.

This chapter will be of interest to assessors and assessment sponsors, and bodies wishing to become certification authorities.

ISO/IEC 15504-3 requires assessors to gain, demonstrate and maintain a level of competence to be recognized as Competent Assessors.

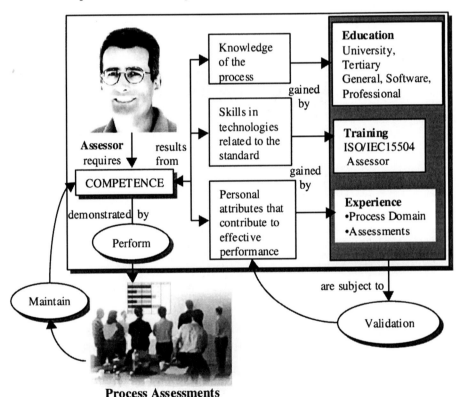

Assessor Competence.

The figure shows the key entities and relationships, which may be articulated as follows:

- The assessor must demonstrates his/her competence to carry out process assessments;
  - The assessor's competence is based upon:
    - their knowledge of the processes;
    - their skills in the principle technologies of ISO/IEC 15504 including the reference model(s), the Process Assessment Models, methods and tools and in the ability to rate processes; and
    - personal attributes which contribute to effective performance, including:
      - professionalism and the ability to develop and maintain the confidence of the assessment participants including creating an open and honest spirit of communication between people so that interviewed personnel will freely discuss issues.
      - the persistent in carrying out the duties that are expected of them and the ability to resolve any conflicts and handle any resistance that they may experience from assessment participants.
      - no conflict of interest in performing the assessment. (one potential conflict may be the inclusion of managers who evaluate the performance of individuals involved in a project being assessed);
      - the ability to assure that the assessed organization has confidence in their judgment and respect for their leadership; and
      - the ability to communicate the findings of the assessments in a clear, non-judgmental style by both verbal and written means.
  - The assessor's knowledge, skills and personal attributes are gained by a combination of education, training and experience.

When the assessor has yet to demonstrate their competence in performing assessments, an alternative is to validate an intending assessor's education, training and experience. This is a precursor to becoming a provisional assessor.

ISO/IEC TR 15504-6:1998 Annexes E and F provide mechanisms for demonstrating competence and mechanisms for validating education, training and experience respectively. These annexes were created in order to set up an assessor registration and certification scheme, but provide can be used by any organization as the basis for selection of persons to be competent assessors.

8.1   Gaining and maintaining competence

Step 1 – Becoming a provisional assessor

A person wishing to perform assessments, must first become a provisional assessor. The provisional assessor must have the required levels of education, training and experience, but may not necessarily have participated in assessments conducted according to the provisions of ISO/IEC 15504.

A provisional assessor should be trained and experienced in the process as well as in process assessment or quality assessment. As ISO/IEC 15504 has several Process Reference Model s describing the process (for example, ISO/IEC 12207, ISO/IEC 15288), it is necessary for the provisional assessor to have competence in at least one of the key process areas in one process model (for example software engineering in accordance with ISO/IEC 12207) that he/she will assess. An assessor should be familiar with software development and maintenance including various life cycle models such as Iterative, Waterfall or Rapid Prototyping. In addition, an assessor should show an understanding of the activities required to support the software process, methods and tools, including when and how they should be applied according to the development model chosen within the application domain in which the assessor is experienced. Lastly, an assessor should be familiar with a range of relevant software engineering standards.

Assessors should demonstrate competence in aspects of the assessment pertaining to ISO/IEC 15504, particularly the core aspects included in parts 2 to 5.

- Overview of the framework for process assessment..
- The process assessment architecture.
- Performing process assessment.
- Conformant assessment models.
- Relevant software standards.

A provisional assessor should also have evidence of an acceptable level of education, both general education (for example covering management) and system/software related education (for example, systems engineering, software engineering, computer science). When validating the provisional assessor's competence it is important to consider the general and specific education together with a combination of training and experience in both system/software development activities and process assessments.

In order to be familiar with software development and maintenance processes, the assessor should have been trained, or have documented experience, in all the processes in the Engineering (ENG) process category. Project management or technical leadership training provides a background in the Customer Supplier (CUS) and the Organizational (ORG) process categories. Assessors need not have been trained in each process in the two process categories, but should be familiar and conversant with the topics. Assessors should have extensive training in at least one of the processes in these two process categories.

Acceptable levels of education may comprise one or more of the following:

- Tertiary courses offered by a college or university (Bachelor Degree or higher);
- Professional courses organized by recognized local or international bodies;
- Vendor sponsored courses; and
- Employer sponsored courses.

When validating the assessor's education, training and experience, consider the balance of general and specific education in terms of:

- Duration: The amount of time the assessor has spent in a particular process category.
- Range: The assessor's breadth of exposure to the process categories.
- Depth: The level of specialization.
- Responsibility: The extent to which an assessor has held responsibility in terms of both range and depth.
- Currency: How recent is the assessor's education, training and experience, and the extent to which the assessor's knowledge and skill have been updated.

Assessors should maintain documented evidence of their education in terms of certificates and course outlines for validation. The following levels of educational achievement may be considered as appropriate in the categories of general education and software education.

For provisional assessors sufficient education and training may be used as an alternative to demonstration of competence.

The provisional assessor must undergo assessor training on ISO/IEC 15504 (and/or any accepted conformant assessment standard such as CMMI® or Automotive SPICE). The training may be:

- Training provided by recognized local or international bodies (for example, the standards organizations); or
- Training provided by vendors and trainers based on the guidance in this part of the International Standard.

There is an interaction between experience and training: training is normally of limited duration and alone is insufficient to achieve full competence in a given process area or to provide the broad experience level needed for assessments. Practical experience is needed to understand detailed application of a process. In addition, experience covering several processes is needed to understand interactions and process interfaces.

Experiences in different roles will aid the overall ability to assess processes as many processes (and their interfaces) are seen from various perspectives. For example, team leaders or project managers may have had contact with software configuration management and software quality assurance functions. The experience gained may help understanding of a number of process categories in any particular assessment.

Due to the nature of ISO/IEC 15504 process assessment where both base practices of a process (process dimension) and management/generic practices (capability dimension) have to be assessed together, recent graduates, or individuals who have spent their entire working lives in a single process category, are unlikely to have accumulated sufficiently broad experience.

Acceptable levels of experience may comprise one or more of the following:

- Direct practical experience in specialist areas such as software engineering, software development/maintenance, software quality, or quality assurance (requires more than just documentation development).
- Management experience overseeing software specialist areas such as software engineering, software development/maintenance, software quality or quality assurance (especially project management experience).
- Participation in customer and supplier interaction activities within an organizational unit with a recognized quality management system (for example, ISO 9000 where the processes will have been documented).

The provision of customer references aids verification of the person's experience.

- Development of plans and the measurement of performance against these plans covering the support process category. Relevant experience includes developing project or user documentation. Assessors should be able to demonstrate familiarity with software quality assurance and quality management systems.
- Experience as managers, consultants or assessors involved in the processes in the organization process category.

In lieu of personal experience, the teaching of the particular subject at a suitable level may suffice, but would require critical examination of the content of course taught.

Step 2 - Becoming a competent assessor

To become a competent assessor, the provisional assessor should have participated in assessments conducted according to the provisions of the International Standard. It is recommended that:

- The provisional assessor participate in at least two assessments led by a competent assessor; or
- The provisional assessor participate in at least one assessment and as an observer in three assessments led by a competent assessor; and
- The provisional assessor leads at least one process assessment under the supervision of a competent assessor.

Competent assessors should maintain a record of ongoing professional activities to demonstrate continuing competencies of skills, knowledge and training.

Step 3 - Maintenance of competence

Competent assessors should maintain their competence by updating their knowledge and skills as well as performing further process assessments, including:

- on the job experience as a competent assessor;
- attending professional seminars;
- giving presentations;
- teaching or developing courses;
- engaging in professional association activities;
- publishing articles or books;

- self training or education using this International Standard; and
- active involvement or leadership in the organizational unit's improvement teams.

ISO/IEC 15504 Specific Training

Provisional assessors need to have received training in ISO/IEC 15504. Several such courses are available[40], and cover the following subjects.

- Background
- Architecture and principles
- The component parts of ISO/IEC 15504
- Vocabulary and definitions
- Comparison of ISO/IEC 15504 with other standards and/or methodologies
- Assessment vs. auditing
- How to use the parts of ISO/IEC 15504

The process assessment architecture

Based on ISO/IEC 15504-2 : *Performing an assessment*

- The Process Dimension:
  - Life cycle process groupings
  - Process categories
  - Basic and component processes
  - Process purposes
- The Capability Dimension:
  - Capability levels
  - Process attributes
- Rating processes and the process Capability Level model
- Requirements for conformant models
- How to use ISO/IEC 15504-2.

Process Assessment

Based on ISO/IEC 15504-3: *Guidance to performing assessments*

- Defining the assessment input
- Responsibilities

---

[40] Course providers include SYNSPACE AG in Europe, Software Quality Institute in Australia,

- The assessment process:
  - Planning
  - Data validation
  - Process rating
  - Reporting
- Recording the assessment output
- Selection and use of a documented assessment process:
  - Using indicators
- Selection and use of assessment instruments and tools
- How to use ISO/IEC 15504-3

Conformant models for assessment

Based on ISO/IEC 15504-2: *Performing an Assessment*; ISO/IEC 15504-3: *Guide to performing assessments*; and ISO/IEC 15504-5: *An exemplar Process Assessment Model* , or any other conformant model.

- The purpose of an assessment model
- Compatibility with the reference model:
  - Purpose and scope of the assessment model
  - Model elements and indicators
  - Mapping the assessment model to the reference model
  - Translating assessment results to process profiles
- Selection and use of a conformant model in assessments

The training should include a case study process assessment that the course participants use to perform a practice assessment.

Maintenance of records

ISO/IEC 15504 specifies a set of records that should be maintained by all assessors and intending assessors:

- educational certificates and course outlines;
- training records describing training courses attended, hours involved, date place and details of the training provider, and specifically verified records of attending training course(s) in the standard (a certificate from the training provider is sufficient);
- verified records of experience in the process domain(s) relevant to process assessment, using the process categories from the Process Reference Model (for example, Engineering); an overview of

involvement in different processes within each category and the level of involvement (for example as a trainee, practitioner, manager, assessor); period or dates of involvement; and verification (by manager, referee);

- verified records of participation in assessments conducted according to the provisions of the standard (by team leader/competent assessor/sponsor);

- verified assessment logs listing date and overview of each assessment (purpose, scope, team size, personal role), duration, process categories assessed; and

- logs of professional activities listing dates, professional activity description, location and duration.

It should be noted that all training, experience and assessments records should be verified by a competent authority.

It is suggested by the author that when a person is not performing assessments on a regular basis, the simplest solution to maintaining records is to keep the details as a specific extension of a personal Curriculum Vitae.

The following template is suggested in ISO/IEC TR 15504-6:1998 and may be used to record an assessor's participation as a provisional assessor or as an observer in assessments conducted according to the provisions of ISO/IEC TR 15504.

Record of Participation

Name of the person:
Date:
No. of days for the assessment:
Scope of the assessment:
Process categories/areas assessed by the person:
Organization/Organizational unit:
Effective Communications:
Were the discussions with the customer reasonable?
Was a satisfactory understanding of ISO/IEC TR 15504 shown?
Was the inter team relationship satisfactory?
Judgment and Leadership:
Were the assessment activities completed in a timely manner?
Were the interviews conducted satisfactorily?
Integrity:
Reasonable sample taken?
Range of activity satisfactory?
Depth of questioning satisfactory?
Review of results consistent?
Rapport :
Communication - telling the good and bad news:
Review of the programme:
Conduct:
Team Management:
Comments: (on Diplomacy, Discretion, Persistence and Resistance handling ability)
Performance: Acceptable/More Experience Required/Not acceptable
Name and signature of assessment sponsor/competent assessor/ team leader:

The involvement in assessments should be verified by the sponsor of the assessment, a competent assessor or the assessment team leader. Each assessment is recorded in a format similar to the one below.

Note: ISO/IEC TR 15504-6:1998 Guide to qualification of Assessors, provides a detailed description of the requirements and example records and logs and should be consulted if further information is required.

8.2   Assessor registration and certification

ISO/IEC TR 15504:1998 proposed an assessor registration system together with validation that assessors met the requirements to be competent assessors.

ISO/IEC TR 15504-6:1998 suggest a mechanism for assessing the education training and experience of a (potential) assessor. Annex G provides checklists that can be used to validate the education, training, experience of (potential) assessors for software process knowledge, assessment technologies (of ISO/IEC 15504) and personal qualities. These are then summed in an Assessor Certification Criteria scoring table.

Validation of the software process competence

Items to validate	Adequacy				Items to Examine	Notes and Commentary
	F	P	N	U		
Software process						
a. Education					• Education accreditation • Degree earned • Number of credit hours • Subject studied	Base or higher degree in a software related discipline preferred.
b. Training					• Training supplier • Type (e.g. instructor led) • Classroom hours • Subject matter • Other assessment models	Training and education alone are not sufficient to become a competent assessor.
c. Experience					• Covers assessment scope • Expertise in at least one process. • Business domain • Application domain • Process variants, if applicable • Other assessment accreditation • Level of responsibility attained	Experience in specific process categories or processes, which are applicable to the assessment scope, should be indicated.

Validation of assessment technology competence

Items to validate	Adequate				Items to Examine	Notes and Commentary
	F	P	N	U		
Assessment Technology						
a. Education					• Educational Institution • Degrees or certificate earned • Classroom hours	Formal education may be used to gain understanding of ISO/IEC TR 15504. Education alone is insufficient to become a competent assessor.

b. Training					• Trainer credentials
					• Type (video, instructor led etc.)
					• Coverage (clause 6.3):
					  – Components of the standard
					  – The reference model
					  – Compatible models
					  – Performing assessments
						Training may be used to obtain knowledge of the components of ISO/IEC TR 15504; assessment methodologies; managing and conducting an assessment.

c. Experience					• Previous assessments conducted • Previous assessment and assessor evaluations • Assessment methodologies used • Satisfied the basis to be a provisional assessor • Creating an assessment methodology, if applicable • Assessment tools and methodologies	This validates the set of experience for consideration of an assessor as a competent assessor.

Validation of personal attributes

Items to validate	Adequate				Items to Examine	Notes and Commentary
	F	P	N	U		
Personal Attributes						
a. Education					• Education accreditation • Degree earned • Number of credit hours	Formal education may include courses in ethics or business philosophy.

b. Training					• Training supplier
					• Type (e.g. instructor led)
					• Classroom hours
					• Subject matter:
					  – Total Quality Management
					  – effective meetings
					  – team building
					  – communication skills
					  – change management
						Look for training completion, understanding and application of principles.

c. Experience					Assessment evaluations Presentations Writing skills Leading change (self-assessments)	Experience is the most reliable indication that an individual possesses the personal attributes desired of an assessor.

Adequacy key:

F: Fully adequate: The information submitted clearly demonstrates that the assessor has the competence in the specific area to successfully perform assessments conducted according to the provisions of ISO/IEC TR 15504.

P: Partially adequate: The information submitted indicates that the assessor has at least some of knowledge and skills necessary to successfully perform assessments conducted according to the provisions of ISO/IEC TR 15504. Additional information may be requested. Alternately, the composition of the assessment team may be altered to include individuals whose knowledge and skills can augment those of the assessor.

N: Not adequate: The information submitted clearly indicates that the assessor does not possess the knowledge and skills in the specific areas to successfully perform assessments conducted according to the provisions of ISO/IEC TR 15504.

U: Unknown: The information submitted does not address the specific knowledge, skills and experience outlined in this part of ISO/IEC TR 15504. Additional information may be needed before a determination can be made.

Assessor Certification Criteria Scoring

Assessor Certification Criteria	Points
Education (Maximum = 4, Minimum = 2)	
Degree or equivalent level of education in any discipline.	1
Any formal course in the Software Process, Computer Science, Software Development, Software Engineering, or Software Quality	1
Degree or equivalent level of education in the Software Process, Computer Science, Software Development, Software Engineering, or Software Quality	2
Assessor education in terms of a national or an international scheme.	2
Training (Maximum = 5, Minimum =3)	
Customer/Supplier process category [CUS]	1
Engineering process category [ENG]	1
Management process category [MAN]	1
Support process category [SUP]	1
Organization process category [ORG]	1
Training based on ISO/IEC TR 15504	3
Experience (Maximum = 5, Minimum = 3)	
Customer/Supplier process category [CUS]	1
Engineering process category [ENG]	1
Management process category [MAN]	1
Support process category [SUP]	1
Organization process category [ORG]	1
Scoring: 9 or above: suitable to be certified; 5 to 8: more education, training & experience needed; below 5: not suitable at present.	

The Software Quality Institute [60] in Australia maintains a register of assessors trained to the standard called the SQI Assessor Programme. This scheme is open to assessors trained through the Software Quality Institute.

The International Assessor Certification Scheme[61] has commenced a registration and certification scheme for assessors. This has replaced the former Certified SPICE Assessor programme.

The International Assessor Certification Scheme operates under the ISO criteria for certification of personnel and is administered by the ASQF (Arbeitskreis Software Qualität Franken, based in Germany). The ASQF ensures that the scheme meets the needs of current method providers, and ensures a harmonisation route for the certification of assessors worldwide. An international Governing Board and Advisory Board will manage the scheme under the general regulations for Certification Schemes, and will include international representation from Certification Bodies, training bodies, assessors, professional bodies, industry representatives and The SPICE User Group.

The scheme provides for three levels of assessor competence: provisional assessor, assessor, and principal assessor. In addition, it will certify training courses and providers. The scheme will address the specific needs of assessors operating in domains such as space, automotive and medical software.

The information required for an assessor to be certified is as explained in the preceding sections of this chapter; including the requirements for education, training, work and assessment experience, and continuing professional development. The logs and forms required are available on the website and are derived from ISO/IEC TR 15504-6:1998.

The scheme details are available at www.int-acs.org .

8.3   Assembling assessment teams

When a sponsor wishes to perform an assessment, he/she should choose an assessment team leader who:

- is a good team builder and leader;
- understands the process domain to be assessed in order to both be able to assess the processes and be able to select team members who complement his/her process and assessment skills.

The appointed assessment team leader needs to define the assessment scope in response to the sponsor's assessment purpose. Refer to the Generic Assessment Procedure in chapter 7 for detailed activities.

When the assessment scope and organizational units to be assessed are known, the appointed assessment team leader should form a team that will:

- cover the processes to be assessed;
- have at least one competent assessor;
- preferably have at least one additional assessor;
- have a balanced set of skills, experience and organizational/process knowledge; and
- preferably include observers.

Therefore, the team leader may need to interview potential team members and/or refer to their assessor records to determine the most suitable team members.

The other assessors should preferably be at least Provisional Assessors but for internal process improvement assessment it is also useful to include persons from the assessed organizational unit to improve commitment, understanding and manage any cultural issues (observers) and/or those who will implement the assessment results.

If absolutely necessary, the team composition may change when covering many organizational unit processes, in this situation the team leader should consider each grouping as a separate team and note which persons were involved in assessing specific processes.

9   SW CMM AND CMMI RELATIONSHIP TO ISO/IEC TR 15504

In this chapter, I describe some of the most important features of the Software Engineering Institute Capability Maturity Models because they are popular Process Assessment Models. I also compare them to ISO/IEC 15504.

Most of the material presented in the following sections of this chapter is based upon information from two sources. The majority of the material comes from an evaluation of CMMI® by Terry Rout, Angela Tuffley and Brent Cahill of the Software Quality Institute in 2002 [62].

A supplementary comparison between CMMI® and ISO/IEC TR 15504-5 comes from Christian Steinmann of HM&S [63]. This latter comparison is a high level overview at a Process Assessment Model level using the older technical report version of the standard. It is not a mapping but informational.

The author has added textual description, and added or modified some tables and figures to improve the explanations.

This chapter will be of interest to anyone wishing to decide which assessment model and method to choose, particularly quality professionals.

The Software Engineering Institute has developed maturity models for software development since 1989, based upon the needs of the United States Department of Defence. The first major published model was the Capability Maturity Model for Software (SW CMM®).

The Software Engineering Institute has researched the relationship between capability, process performance and maturity and provides the following definitions in the SW CMM® V1.1 [64].

Software process capability describes the range of **expected** results that can be achieved by following a software process. The software process capability of an organization provides one means of predicting the most likely outcomes to be expected from the next software project the organization undertakes.

Software process performance represents the **actual** results achieved by following a software process. Thus, software process performance focuses on the results achieved, while software process capability focuses on results expected. Based on the attributes of a specific project and the context within which it is conducted, the actual performance of the project may not reflect the full process capability of the organization; i.e., the capability of the project is constrained by its environment.

Software process maturity is the extent to which a specific process is explicitly defined, managed, measured, controlled, and effective.

Note: In the CMMI® this has been replaced by a definition for organisational maturity.

In the same introduction, the Software Engineering Institute states that it defined the staged structure of the SW CMM® based on principles of product quality that have been formulated over the last sixty years, and based in part upon the work of Walter Shewart (principles of statistical quality control), W. Edwards Deming and Joseph Juran. The Software Engineering Institute adopted the principles into a maturity framework that establishes a project management and engineering foundation for quantitative control of the software process, which is the basis for continuous process improvement.

This staged representation for implementation of processes has specific processes needed to achieve a particular maturity level.

SW CMM® Maturity Levels and Processes

Maturity		Processes – additional processes per maturity level
Maturity Level 1	Initial	no processes, chaos
Maturity Level 2	Managed	Requirements Management, Software Project Planning, Software Project Tracking and Oversight, Software Subcontract Management, Software Quality Assurance, Software Configuration Management.
Maturity Level 3	Defined	Organizational Process Focus, Organizational Process Definition, Training Program, Integrated Software Management, Software Product Engineering, Intergroup Coordination, Peer Reviews
Maturity Level 4	Quantitatively Managed	Process Measurement and Analysis, Quality Management
Maturity Level 5	Optimising	Defect Prevention, Technology Innovation, Process Change Management

The staged representation was a model that software organizations found easy to understand due to the explicit processes to be deployed at each maturity level. Many organizations in North America and India (and to a much lesser extent in Europe and Asia) have adopted SW CMM® V1.1[41] as

---

[41] The Software Engineering Institute released a draft version 2 of the SW CMM® in 1997. It proposed new processes/practices and revisions to existing processes/practices with greater coverage of ISO/IEC 15504 processes. However due to the nature of changes in the model,

their preferred assessment framework and their preferred process deployment approach.

The model reflects its original client and industrial organizations; covering large and/or complex software projects in organizations delivering mission critical systems (initially for military purposes). It is by nature prescriptive in terms of assessing the staging of process implementation.

An advantage of a SW CMM® assessment is that it gives a single Maturity Level (ML) for the assessed organization. The result can be expressed as a single number and this is attractive to clients and organizations, who do not wish to investigate in detail the meaning of the result (for example, to which domains it applies).

Due to the staged nature of process deployment, it is restrictive in terms of adaptability/flexibility; an organization cannot decide to adopt processes in a different order to that described in the model without ramifications on the assessed maturity level. While it does suit many organizations to use this staged approach, they need to be aware that it is not conformant to the international standard ISO/IEC 15504, which has ramifications on comparability of assessment results with other assessment methods.

The Software Engineering Institute had in parallel to the later development of the SW CMM® , also developed a System Engineering CMM® , an Integrated Product Development CMM® and a Software Acquisition CMM® . It became apparent that there were inconsistencies, overlaps and duplications between the models. They made a decision that the most important aspects of these models should be integrated. This decision resulted in creation of the Capability Maturity Model Integration (CMMI®) [65] which has been developed in several variants (the Systems Engineering, Software Engineering, and Integrated Product and Process Development variant is referred to here as it is the most comprehensive variant).

The CMMI® was conceived to both comply with the emerging ISO/IEC 15504 standard and as a staged representation, thereby maintaining representational consistency with SW CMM® . The staged representation of CMMI® can be shown in a simplified form as maturity levels that indicate Process Areas (the set of processes in the table above). Each Process Area has Specific Goals achieved by Specific Practices, and Generic Goals that

the USAF sponsor's desire to integrate various CMMs and achieve conformance to ISO/IEC 15504 it was decided not to adopt it. Hence SW CMM® V1.1 is still in use. There is planning in the SEI to retire the SW CMM® with a target date by the end of 2005, so users will need to consider migrating to CMMI® or ISO/IEC 15504.

are achieved by Generic Practices. The Generic Goals can be summarized as: commitment to perform; ability to perform; directing implementation; and verifying implementation.

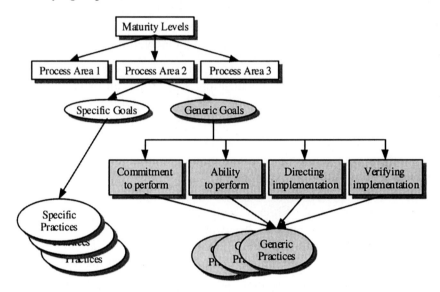

Staged representation of CMMI.

CMMI® Maturity Levels and processes

Maturity		CMMI Processes – additional processes per maturity level
Maturity Level 1	Initial	No processes
Maturity Level 2	Managed	Requirements Management, Measurement and Analysis, Project Monitoring and Control, Project Planning, Process and Product Quality Assurance, Supplier Agreement Management, Configuration Management
Maturity Level 3	Defined	Decision Analysis and Resolution, Product Integration, Requirements Development, Technical Solution, Verification, Validation, Organizational Process Focus, Organizational Process definition, Integrated Project Management, Risk management, Organizational training, Integrated Teaming, Organizational Environment for Integration
Maturity Level 4	Quantitative ly Managed	Organizational Process Performance, Quantitative Project management
Maturity Level 5	Optimising	Organizational Innovation and Deployment, Causal Analysis and resolution

Source: Software Quality Institute. CMMI Evaluation. Capability Maturity Model Integration Mapping to ISO/IEC TR 15504-2:1998

It is apparent that the processes in the CMMI® are different to those in the SW CMM® . The system engineering aspects are the major source of additional processes, but there are also additional advanced support and organizational processes. The United States Air Force Software Technology Support Center has prepared a mapping between the SW CMM® V1.1 and the CMMI® V0.2. I will not focus on the mapping here but provide the reference [66] for those interested in further research.

While developing the CMMI® , the Software Engineering Institute has been involved in the development of ISO/IEC 15504 and therefore has also created a continuous representation of CMMI® . This representation is still structurally different to the Process Assessment Model in ISO/IEC 15504-5 but is intended to be mapped to a conformant Process Reference Model.

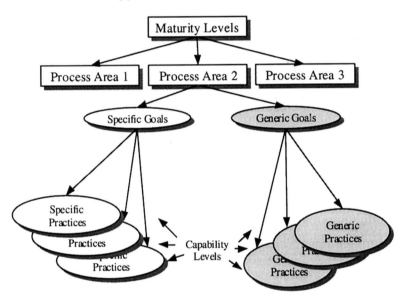

Continuous representation of CMMI.

As will be described in the following section, the mapping of the continuous representation from CMMI® to ISO/IEC 15504 is not straightforward.

9.1   Scope of CMMI® compared to ISO/IEC 15504

The first important aspect when comparing the scope of CMMI® and ISO/IEC 15504 is that the technical solution processes in the CMMI® cover both Systems Engineering and Software Engineering. This broader scope of the CMMI® is an important reason for large system integrators and acquirers to consider using it. If ISO/IEC 15504 uses ISO/IEC 15288 and ISO/IEC 12207 to create a composite process dimension, then a similar (but not the same) scope covering systems and software engineering is possible.

CMMI® Continuous Representation Model Process Dimension

The CMMI® Continuous Representation Model is the simpler representation to compare to ISO/IEC 15504. In this representation, there are four major process categories as shown in the following table.

CMMI® continuous representation model process areas

Category	Process Area
Process Management	Organizational Process Focus Organizational Process Definition Organizational Training Organizational Process Performance Organizational Innovation and Deployment
Project Management	Project Planning Project Monitoring and Control Supplier Agreement Monitoring Integrated Project management Integrated Teaming Risk Management Quantitative Project management
Engineering	Requirements Management Requirements Development Technical Solution Product Integration Verification Validation
Support	Configuration Management Process and Product Quality Assurance Measurement and Analysis Causal Analysis and Resolution Decision Analysis and Resolution Organizational Environment for Integration

CMMI® as defined in the Continuous Representation Model has some differences to the process dimension of ISO/IEC 15504-5.

The CMMI® technical solution processes can be mapped in part against ISO/IEC 15288 as a Systems Engineering Process Reference Model , although the detailed mapping is not yet completed.

CMMI® addresses all of the processes of ISO/IEC 15504 Process Reference Model, with the exception of the processes specifically identified below.

- CUS.4 - Operation Process

- MAN.1 - Management Process
- ORG.1 - Process Alignment Process

The following processes are not completely addressed in the CMMI® V1.1:

- CUS.2 - Supply Process
- ENG.2 - Software Maintenance Process
- ORG.3 - Human Resource Management Process
- ORG.6 - Reuse Process

In addition, CMMI® does not have a Customer-Supplier process category but contains the Customer - Supplier processes and sub-processes in the Engineering and Project Management Process Areas. Establish Supplier Agreements and Satisfy Supplier Agreements in the Project Management Area partially cover the Acquisition Process and the Supply Process in ISO/IEC 15504. Manage Requirements in the Engineering Process Area provides equivalent coverage to Requirements Elicitation in ISO/IEC 15504.

The CMMI® uses a cyclic model for the Engineering processes, while ISO/IEC 15504 does not specify cyclic or phased models (either may be valid).

CMMI® Continuous Representation Model Capability Dimension

In CMMI® the capability dimension is represented by the Generic Practices, plus some process areas.

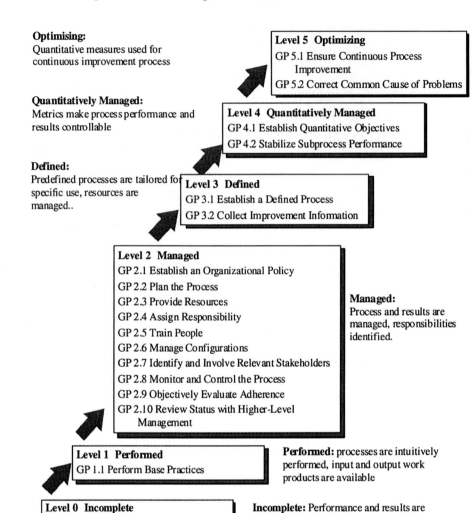

Optimising:
Quantitative measures used for continuous improvement process

Level 5 Optimizing
GP 5.1 Ensure Continuous Process Improvement
GP 5.2 Correct Common Cause of Problems

Quantitatively Managed:
Metrics make process performance and results controllable

Level 4 Quantitatively Managed
GP 4.1 Establish Quantitative Objectives
GP 4.2 Stabilize Subprocess Performance

Defined:
Predefined processes are tailored for specific use, resources are managed..

Level 3 Defined
GP 3.1 Establish a Defined Process
GP 3.2 Collect Improvement Information

Level 2 Managed
GP 2.1 Establish an Organizational Policy
GP 2.2 Plan the Process
GP 2.3 Provide Resources
GP 2.4 Assign Responsibility
GP 2.5 Train People
GP 2.6 Manage Configurations
GP 2.7 Identify and Involve Relevant Stakeholders
GP 2.8 Monitor and Control the Process
GP 2.9 Objectively Evaluate Adherence
GP 2.10 Review Status with Higher-Level Management

Managed:
Process and results are managed, responsibilities identified.

Level 1 Performed
GP 1.1 Perform Base Practices

Performed: processes are intuitively performed, input and output work products are available

Level 0 Incomplete

Incomplete: Performance and results are incomplete, chaotic processes

CMMI® Capability Dimension.

The Generic Practices in CMMI® do not fully cover all of the Process Attributes in CMMI. As stated in the SQI report: *"Two of the process attributes of the Capability Dimension of ISO 15504-2 are not addressed in any of the Generic Practices of the CMMI Continuous Model. These are PA3.2 - Process Resource Attribute, and PA5.1 - Process Change Attribute"*

The SQI mapping shows that these two process attributes are addressed in process areas of the CMMI so that coverage of all attributes is achieved.

Process attribute PA 3.2 is addressed in Organizational Process Focus, Organizational Process Deployment, Organizational Training and Integrated Project Management process areas. This means that any assessment to cover ISO/IEC 15504 Capability Level 3 would require these process areas to be in the assessment scope in order to be able to translate the results.

Process attribute PA 5.1 is addressed in Organizational Innovation and Deployment and Causal Analysis and Resolution. This means that any assessment to cover ISO/IEC 15504 Capability Level 5 would require these process areas to be in the assessment scope in order to be able to translate the results.

On the other hand, most of the Work Product Management aspects in process attribute PA 2.2 are not explicitly addressed in the CMMI®.

The need to assess process areas in order to obtain Capability Level ratings is a possible disincentive to use the continuous representation. The increase in assessment scope also works against making narrower scope assessments of other process areas if wishing to determine higher Capability Levels for those processes.

While I am unaware of organizations using CMMI® and translating the results into ISO/IEC 15504 Capability Levels (rather than using the staged model maturity levels), this may not remain an academic concern as ISO/IEC 15504 gains wider acceptance.

CMMI® areas outside the scope of ISO/IEC TR 15504-2

Some elements in the CMMI® Continuous Representation - the whole of the Decision Analysis and Resolution process area (advanced support process) and parts of the Technical Solution process area (particularly systems engineering aspects) are outside the scope of the earlier technical report version of ISO/IEC TR 15504-2.

The technical solution processes can be mapped in part against ISO/IEC 15288 as a Systems Engineering Process Reference Model , although the detailed mapping is not yet completed.

Issues concerning CMMI® mapping and translation of assessment results

As there is no precise or simple one-to-one mapping between CMMI® and ISO/IEC 15504, the assessor needs to take care when translating the results of a CMMI® conformant assessment to the standard. Some processes, for example SUP.1 - Documentation Process and ORG.4 - Infrastructure

Process, are addressed rather generally in CMMI® over a wide range of process areas. It is probable in an assessment that insufficient data would be collected to permit rating of these processes using any feasible translation mechanism.

In some process areas, especially for Requirements Management and Integrated Project Management, the mapping (even at the sub-practice level) is to a wide range of process outcomes. This will lead to significant problems in attempting to use observations recorded in these process areas in any translation. It also implies that the definition of these process areas is inadequately detailed; for example, Requirements Management has only one Specific Goal and four Specific Practices.

There are concerns relating to the use of advanced practices in the CMMI® Continuous Model. In most instances, where an Advanced Practice represented an increased level of capability of an equivalent basic practice, the only additional material relates to mappings to process attributes (as expected). In some cases, however, there are additional process outcomes included in the mappings to the advanced practice, and in one case (Technical Solution Sub Practices 1.1-1, 1.1-2) there is one outcome addressed in the Basic Practice that is not covered in the Advanced Practice. This is an issue with the overall design and integrity of the Model.

CMMI® Staged Representation Model

The primary difficulty in mapping the CMMI® Staged Representation Model is in relation to the Capability Dimension as only Level 2 and Level 3 Generic Practices are explicitly identified, and these are established within each Process Area, rather than being seen as common capabilities across all areas. In undertaking this mapping, therefore, the mapping to the Generic Practices previously established was taken, and in each process area the relevant Process Attributes were seen as applying to the cited Processes. In describing this coverage, the following guidelines were adopted:

- If all outcomes of a Process are addressed in the Basic Practices of the Process Area, the attributes derived from the relevant Generic Practices are seen as applying *completely* to that Process.
- If more than one outcome of a Process is addressed, the attributes derived from the relevant Generic Practices are seen as applying *partially* to that Process.

- If only a single outcome of a Process is addressed, the attributes derived from the relevant Generic Practices are regarded as applying *slightly* to that Process.

Within a single Process Area, therefore, the Performance Management, Work Product Management and Process Definition process attributes (for Process Areas at Level 3 and higher) can be seen as applying to different extents to different Processes in the Reference Model.

CMMI® Staged Representation Maturity Level processes and ISO/IEC TR 15504

Maturity		CMMI Processes – additional processes per maturity level	ISO/IEC 15504 mapped processes
Maturity Level 1	Initial	No processes	No direct equivalent
Maturity Level 2	Managed	Requirements Management, Measurement and Analysis, Project Monitoring and Control, Project Planning, Process and Product Quality Assurance, Supplier Agreement Management, Configuration Management	Supplier Selection Process, Supplier Monitoring Process, Customer Acceptance Process, Documentation Process, Configuration Management Process, Quality Assurance Process, Joint Review Process, Audit Process, Problem Resolution Process, Project Management Process, Measurement Process
Maturity Level 3	Defined	Decision Analysis and Resolution, Product Integration, Requirements Development, Technical Solution, Verification, Validation, Organizational Process Focus, Organizational Process definition, Integrated Project Management, Risk management, Organizational training, Integrated Teaming, Organizational Environment for Integration	Requirements Elicitation Process, System requirements analysis and design process, Software requirements analysis process, Software design process, Software construction process, Software integration process, Software testing process, System integration and testing process, Verification Process, Validation Process, Risk Management, Process establishment process, Process assessment process,
Maturity Level 4	Quantitative ly Managed	Organizational Process Performance, Quantitative Project management	Quality Management, Infrastructure Process.
Maturity Level 5	Optimising	Organizational Innovation and Deployment, Causal Analysis and resolution	Process Improvement process

In terms of the scope of the model, it is clear that the process dimension of the CMMI® Continuous Representation Model covers the same scope as does the Continuous Representation.

Issues and Anomalies of the CMMI® Staged Representation Model

The issues raised in respect of the Continuous Representation apply also to the Staged Representation. In addition, there are concerns relating to the non-inclusion of Generic Practices for Levels 4 and 5 into the relevant Process Areas.

There is also an apparent anomaly in respect of capability of lower-level Process Areas. With the Continuous Representation, Process Areas such as Configuration Management or Requirements Management can be seen as evolving over a full range of capabilities; thus, it is meaningful to speak of a "Level 4 (capability) Requirements Management Process Area".

9.2 Capability Levels and Maturity Levels

The CMMI® Staged Representation Model uses the standard five maturity levels. When the mapping is performed against the ISO/IEC TR 15504-2:1998 equivalent processes (see the earlier table: CMMI® Maturity Level processes and equivalent ISO/IEC TR 15504 processes), the following mapping results.

Maturity Level 2 Equivalent Processes

Maturity Level 2	
CUS.1.2	Supplier Selection Process
CUS.1.3	Supplier Monitoring Process
CUS.1.4	Customer Acceptance Process
SUP.1	Documentation Process
SUP.2	Configuration Management Process
SUP.3	Quality Assurance Process
SUP.6	Joint Review Process
SUP.7	Audit Process
SUP.8	Problem Resolution Process
MAN.2	Project Management Process
ORG.5	Measurement Process
Process Attributes	PA1.1 PA2.1 PA2.2

Maturity Level 3 Equivalent Processes

Maturity Level 3	
CUS.1.2	Supplier Selection Process
CUS.1.3	Supplier Monitoring Process
CUS.1.4	Customer Acceptance Process
CUS.3	Requirements Elicitation Process
ENG.1.1	System requirements analysis and design process
ENG.1.2	Software requirements analysis process
ENG.1.3	Software design process
ENG.1.4	Software construction process
ENG.1.5	Software integration process
ENG.1.6	Software testing process
ENG.1.7	System integration and testing process
SUP.1	Documentation Process
SUP.2	Configuration Management Process
SUP.3	Quality Assurance Process
SUP.4	Verification Process
SUP.5	Validation Process
SUP.6	Joint Review Process
SUP.7	Audit Process
SUP.8	Problem Resolution Process
MAN.2	Project Management Process
MAN.4	Risk Management
ORG.2.1	Process establishment process
ORG.2.2	Process assessment process
ORG.5	Measurement Process
Process Attributes	PA1.1 PA2.1 PA2.2 PA3.1 PA3.2

Maturity Level 4 Equivalent Processes

Maturity Level 4	
CUS.1.2	Supplier Selection Process
CUS.1.3	Supplier Monitoring Process
CUS.1.4	Customer Acceptance Process
CUS.3	Requirements Elicitation Process
ENG.1.1	System requirements analysis and design process
ENG.1.2	Software requirements analysis process
ENG.1.3	Software design process
ENG.1.4	Software construction process
ENG.1.5	Software integration process
ENG.1.6	Software testing process
ENG.1.7	System integration and testing process
SUP.1	Documentation Process
SUP.2	Configuration Management Process
SUP.3	Quality Assurance Process
SUP.4	Verification Process
SUP.5	Validation Process
SUP.6	Joint Review Process
SUP.7	Audit Process
SUP.8	Problem Resolution Process
MAN.2	Project Management Process
MAN.3	Quality Management
MAN.4	Risk Management
ORG.2.1	Process establishment process
ORG.2.2	Process assessment process
ORG.4	Infrastructure process
ORG.5	Measurement Process
Process Attributes	PA1.1 PA2.1 PA2.2 PA3.1 PA3.2 PA4.1 PA4.2

Maturity Level 5 Equivalent Processes

Maturity Level 5	
CUS.1.2	Supplier Selection Process
CUS.1.3	Supplier Monitoring Process
CUS.1.4	Customer Acceptance Process
CUS.3	Requirements Elicitation Process
ENG.1.1	System requirements analysis and design process
ENG.1.2	Software requirements analysis process
ENG.1.3	Software design process
ENG.1.4	Software construction process
ENG.1.5	Software integration process
ENG.1.6	Software testing process
ENG.1.7	System integration and testing process
SUP.1	Documentation Process
SUP.2	Configuration Management Process
SUP.3	Quality Assurance Process
SUP.4	Verification Process
SUP.5	Validation Process
SUP.6	Joint Review Process
SUP.7	Audit Process
SUP.8	Problem Resolution Process
MAN.2	Project Management Process
MAN.3	Quality Management
MAN.4	Risk Management
ORG.2.1	Process establishment process
ORG.2.2	Process assessment process
ORG.2.3	Process improvement process
ORG.4	Infrastructure process
ORG.5	Measurement Process
Process Attributes	PA1.1 PA2.1 PA2.2 PA3.1 PA3.2 PA4.1 PA4.2 PA5.1 PA5.2

The Software Quality Institute determined the equivalent ISO/IEC TR 15504-2:1998 process capability profile by mapping from each of the Specific and Generic Practices for each Process Area in the maturity level to the outcomes and process attributes of the ISO/IEC TR 15504-2:1998 processes.

The tables show what set of processes must be implemented to achieve the equivalent process attributes for each maturity level. It does not mean that each individual process is for example at Capability Level 4 or 5.

To include an ISO/IEC 15504 process in the maturity level profile, all outcomes for an ISO/IEC 15504 process and process attribute achievements need to be covered by the mapping from the Specific Practices and Generic Practices of the CMMI Staged representation. The results of this are presented in the following figure.

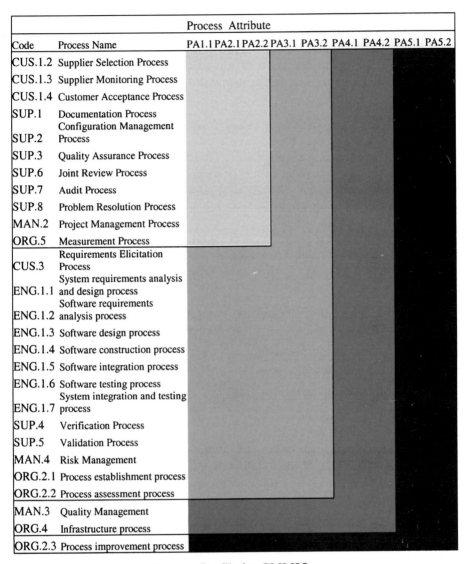

		Process Attribute							
Code	Process Name	PA1.1	PA2.1	PA2.2	PA3.1	PA3.2	PA4.1	PA4.2	PA5.1 PA5.2
CUS.1.2	Supplier Selection Process								
CUS.1.3	Supplier Monitoring Process								
CUS.1.4	Customer Acceptance Process								
SUP.1	Documentation Process								
SUP.2	Configuration Management Process								
SUP.3	Quality Assurance Process								
SUP.6	Joint Review Process								
SUP.7	Audit Process								
SUP.8	Problem Resolution Process								
MAN.2	Project Management Process								
ORG.5	Measurement Process								
CUS.3	Requirements Elicitation Process								
ENG.1.1	System requirements analysis and design process								
ENG.1.2	Software requirements analysis process								
ENG.1.3	Software design process								
ENG.1.4	Software construction process								
ENG.1.5	Software integration process								
ENG.1.6	Software testing process								
ENG.1.7	System integration and testing process								
SUP.4	Verification Process								
SUP.5	Validation Process								
MAN.4	Risk Management								
ORG.2.1	Process establishment process								
ORG.2.2	Process assessment process								
MAN.3	Quality Management								
ORG.4	Infrastructure process								
ORG.2.3	Process improvement process								

Process Profile for CMMI®.

Processes are not listed in the profile until all defined outcomes have been satisfied. In most cases, some outcomes are addressed at lower levels of maturity; however, all outcomes must be satisfied to demonstrate completeness of process performance.

The Software Quality Institute mapping demonstrates that the results of a CMMI® assessment can provide sufficient evidence to support the production of a comprehensive series of ISO/IEC 15504 process profiles.

However, due to the complexity of the mappings, the high level at which ratings are assigned, and the nature of the rating process, it is not seen as feasible, given the current state of knowledge, to develop an automatic translation mechanism. Any mechanism will require additional judgment by the assessment team.

CMMI® GPs and ISO/IEC TR 15504-5:1998 MPs

For readers more familiar with ISO/IEC TR 15504-5, the following diagrams represent a broad comparison of the Generic Practices of CMMI® and the Management Practices of ISO/IEC TR 15504-2:1998 (Source: Christian Steinmann). They are informational and not an assessment model mapping.

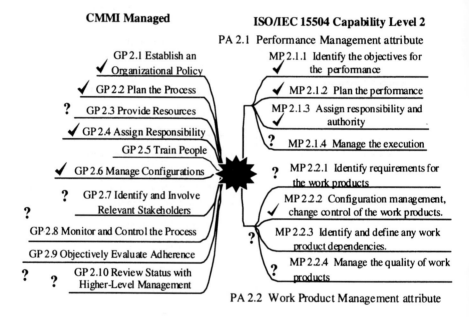

CMMI® and ISO/IEC TR 15504-2:1998 Capability level 2.

At Capability Level 2, the following Generic Practices (GP) and Management Practices (MP) are equivalent: GP 2.1 and MP 2.1.1; GP 2.2 and MP 2.1.2; GP 2.4 and MP 2.1.3; GP 2.6 and MP 2.2.2.

GP 2.3 only partly covers the process attribute PA 3.2 at project level (coverage of PA 3.2 requires several process areas in CMMI® comprising Organizational Process Focus, Organizational Process Deployment, Organizational Training and Integrated Project Management). GP 2.5 covers part of MP3.2.1. GP2.7 is related to MP 2.1.1 and MP 2.1.3. GP 2.8 is related to MP2.1.4 but the former has a process control orientation versus a process performance orientation in MP2.1.4.

Three of the Management Practices in PA 2.2, MP 2.2.1, MP 2.2.3 and MP 2.2.4 are work product focused and not directly covered in the CMMI®, neither by the GPs, nor in the Level 2 Process Areas. This lack of explicit work product management coverage is an area where the two models disagree on a critical aspect of capability.

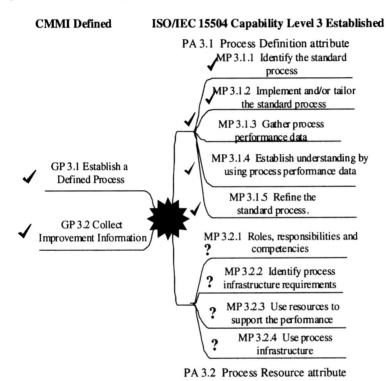

CMMI® and ISO/IEC TR 15504-2:1998 Capability level 3.

At Capability Level 3, the following Generic Practices and Management Practices cover the same purpose: GP 3.1 and MP 3.1.1, MP 3.1.2 and MP 3.1.5; GP 2.2 and MP 3.1.3 and MP 3.1.4. It should be noted that establishing a Standard process is not explicit in the GPs, but is implicit, through the definition of the term "defined process". No Generic Practice at this Capability Level exactly covers the process attribute Process Attribute PA 3.2 (they are covered by CMMI® process areas).

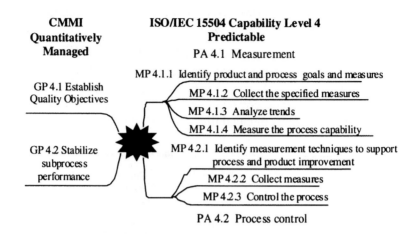

CMMI and **ISO/IEC TR 15504-2:1998 Capability level 4.**

At Capability Level 4, the Generic Practices and Management Practices are related but not equivalent. GP 4.1 sets general quantitative quality objectives for processes, allocated to sub-processes, while PA 4.1 defines more specific steps in quantitative goals and measures. GP 4.2 specifically requires statistical process control of **critical sub-processes** to manage and predict process performance, while PA 4.2 only specifies use of measurement.

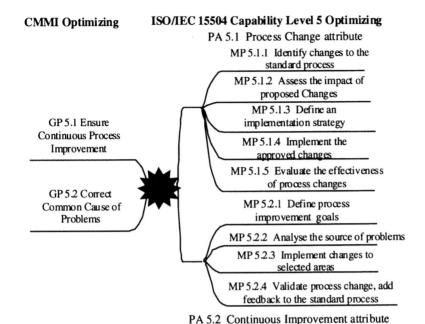

CMMI Optimizing

ISO/IEC 15504 Capability Level 5 Optimizing

PA 5.1 Process Change attribute

MP 5.1.1 Identify changes to the standard process

MP 5.1.2 Assess the impact of proposed Changes

MP 5.1.3 Define an implementation strategy

MP 5.1.4 Implement the approved changes

MP 5.1.5 Evaluate the effectiveness of process changes

GP 5.1 Ensure Continuous Process Improvement

GP 5.2 Correct Common Cause of Problems

MP 5.2.1 Define process improvement goals

MP 5.2.2 Analyse the source of problems

MP 5.2.3 Implement changes to selected areas

MP 5.2.4 Validate process change, add feedback to the standard process

PA 5.2 Continuous Improvement attribute

CMMI® and ISO/IEC TR 15504-2:1998 Capability level 5.

At Capability Level 5, the Generic Practice GP 5.1 and Process Attribute PA 5.2 are equivalent. The Generic Practice GP 5.2 maps to MP 5.2.2, in conjunction with MP 5.2.3. On the other side, PA 5.1 has no direct equivalent GP. PA 5.1 does relates to the Organizational Innovation and Deployment and Causal Analysis and Resolution Process Areas in CMMI®.

In summary, not all aspects of the ISO/IEC TR 15504-2 Process Attributes are contained in the Capability Dimension of the CMMI® continuous representation model. PA3.2 - Process Resource Attribute and PA5.1 - Process Change Attribute are covered by several process areas in the CMMI®. More importantly PA 2.2 is not well covered in the CMMI® and this lack of explicit work product management coverage is an area where the two models disagree on a critical aspect of capability.

One important caveat on the analysis in this section – the revised capability dimension process attributes in ISO/IEC 15504 will lead to revised management practices (or generic practices) in ISO/IEC 15504-5. Some of the anomalous issues covered in this section will no longer be a problem when the new part 5 is issued.

9.3   CMMI® Assessments

There are three defined classes of CMMI® assessments. The SEI document Appraisal Requirements for CMMI [67] provides the details, that are summarised here.

Class A assessments are required in order to derive Maturity levels for an organization, while Class B and Class C are less rigorous and can provide a less time consuming alternative when Maturity levels are not required. SCAMPI (Standard CMMI Appraisal Method for Process Improvement) meets the class A requirements.

The need to be able to achieve translation to ISO/IEC 15504 Process Profiles is only required for Class A methods, and the SEI document specifies some additional requirements to meet this need.

The following table provides a high level summary of the appraisal classes.

Classes of Appraisal Methods

Characteristics	Class A	Class B	Class C
Usage mode	1. Rigorous and in-depth investigation of process(es) 2. Basis for improvement plan	1. Initial (first-time) 2. Incremental (partial) 3. Self-assessment	1. Quick-look 2. Incremental
Principal Outcomes	1. Findings adequate as a basis for process improvement activities 2. Buy-in and ownership of results 3. CMMI measurement framework rating(s) to characterize assessment scope	1. Findings adequate as a basis for process improvement program 2. Buy-in and ownership of results.	1. Findings adequate to expose gaps in implementation of processes
Advantages	Thorough coverage; strengths and weaknesses for each PA investigated; robustness of method with consistent, repeatable results; provides objective view; option of 15504 conformance.	Organization gains insight into own capability; provides a starting point or focuses on areas that need most attention; promotes buy-in.	Inexpensive; short duration; rapid feedback.
Disadvantages	Demands significant resources.	Does not emphasize depth of coverage and rigor and cannot be used for level rating.	1. Provides less buy-in and ownership of results. 2. Not enough depth to fine-tune process improvement plans.
Sponsor	Senior manager of organizational unit.	Any manager sponsoring an SPI program.	Any internal manager.
Team composition	External and internal.	External or internal	External or internal.
Team size	4-10 persons + assessment team leader.	1-6 + assessment team leader.	1-2 + assessment team leader.
Team qualifications	Experienced.	Moderately experienced.	Moderately experienced.
Assessment team leader requirements	Lead assessor.	Lead assessor or person experienced in method.	Person trained in method.

Source: The Software Quality Institute CMMI Evaluation report. See Table 1 and Appendix A of the SEI document: Appraisal Requirements for CMMI for the details used to create this table.

9.4 Use of SW CMM® and CMMI®

CMMI® appeals to the software and systems development community already familiar with the SW CMM® It appeals particularly to government and government contractors (e.g. contractors to the United States and Australian Departments of Defence). Over 50% of assessments were reported for this organization type and they prefer to naturally only have to perform one type of assessment for both their civil and military business areas.

Companies in the defense and aerospace business (especially large prime contractors) have adopted CMMI® as their preferred assessment method.

It also has attracted up until now more medium to large companies than small companies (over 50% of assessments had more than 200 employees within the **area** of the organization appraised).

75% of assessments used the CMMI® Staged Representation Model.

The following results come from the Process Maturity Profile report made by the Software Engineering Institute in September 2003 [68].

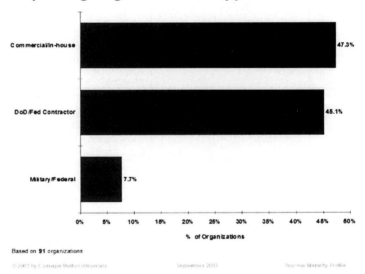

CMMI® assessed organization by type.

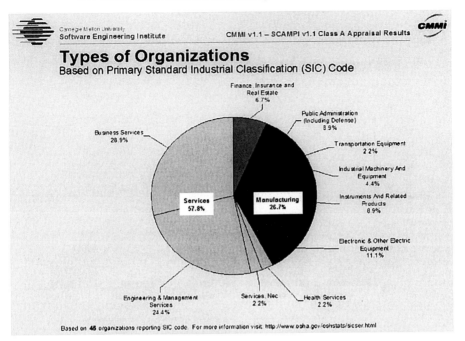

CMMI® assessed organizations by business type.

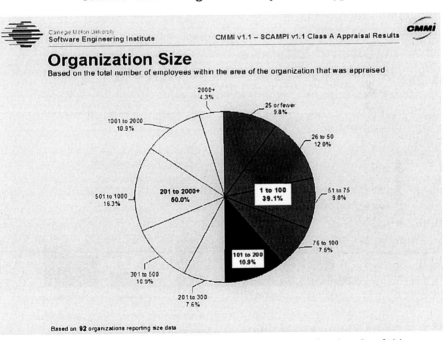

CMMI® assessed organizations by assessed organizational unit(s).

Carnegie Mellon University
Software Engineering Institute   CMMI v1.1 – SCAMPI v1.1 Class A Appraisal Results   *CMMI*

Maturity Profile by Organizational Type

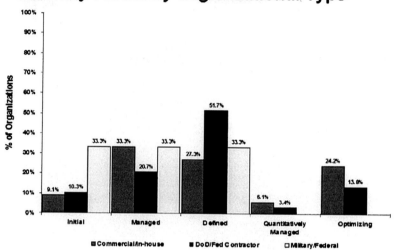

Based on most recent appraisal of **65** organizations reporting organization type and a maturity level rating

CMMI® Maturity Profile – assessment in 2002 to 2003.

Carnegie Mellon University
Software Engineering Institute   CMMI v1.1 – SCAMPI v1.1 Class A Appraisal Results   *CMMI*

USA and Offshore
Summary Organizational Maturity Profiles

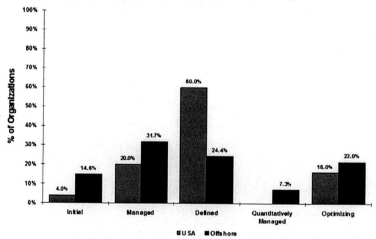

Based on **25** U.S. organizations and **41** offshore organizations reporting their maturity level rating

CMMI® Process Maturity – USA and Offshore.

The SW CMM® is still popular with a wider variety companies due to its relative simplicity and longer history in comparison with CMMI® . There are still many companies involved in developing software for the United States government who are using SW CMM® and this has been promulgated to companies used for outsourcing in countries like India. The following results are taken from the Software Engineering Institute report on Process Maturity Profile for Software CMM [69].

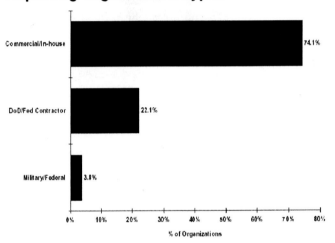

SW CMM® assessed organization types.

Carnegie Mellon University
Software Engineering Institute   Software CMM - CBA IPI and SPA Appraisal Results

Types of Organizations
Based on Primary Standard Industrial Classification (SIC) Code

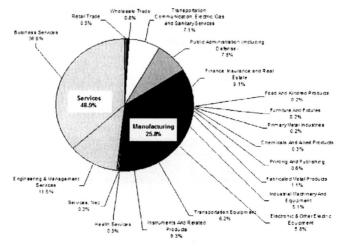

Based on **650** organizations reporting SIC code. For more information visit: http://www.osha.gov/oshstats/sicser.html

SW CMM® assessed organization by business type.

Carnegie Mellon University
Software Engineering Institute   Software CMM - CBA IPI and SPA Appraisal Results

Organization Size
Based on the total number of employees within the area of the organization that was appraised

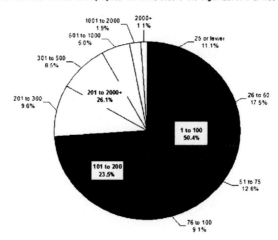

Based on **1,312** organizations reporting size data

SW CMM® assessed organization by size of assessed organizational unit(s).

 Carnege Mellon University
Software Engineering Institute   Software CMM - CBA IPI and SPA Appraisal Results

Maturity Profile by Organization Type

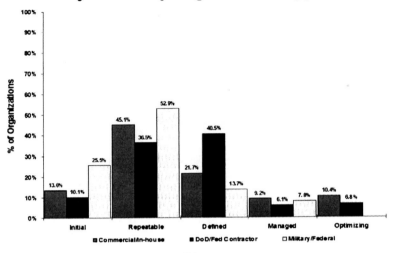

Based on most recent appraisal, since 1999, of 1,342 organizations

SW CMM® Process Maturity.

 Carnege Mellon University
Software Engineering Institute   Software CMM - CBA IPI and SPA Appraisal Results

USA and Offshore
Organization Maturity Profiles

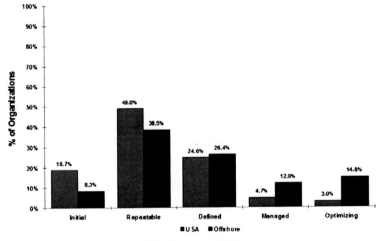

Based on 641 U.S. organizations and 701 offshore organizations

SW CMM® Process Maturity – USA and Offshore.

Time to Move Up

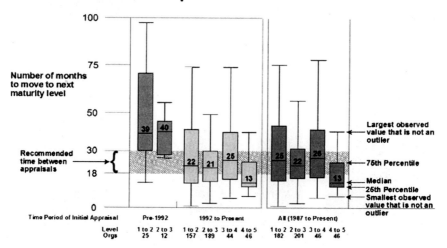

Time period (months) to move up from one maturity level to the next.

The report clearly indicates that SW CMM® and CMMI® Process Maturity have been increasing. It is also evident that SW CMM® is still more popular than CMMI®.

9.5 Advantages and Disadvantages

CMMI® is about 700 pages and the Standard CMMI Appraisal Method for Process Improvement (SCAMPI^{SM}) is 245 pages long.

CMMI® provides a much greater level of details for the processes. It describes sub-practices, which are actually implementation hints under the specific practices.

CMMI® defines typical work products on the level of specific practices and these are not always consistent with the same or similar work products in other practices.

CMMI® Staged Representation Model suits larger process-centric organizations, but often does not suit smaller organizations. It is also requires more effort with organizations that allow high project independence in terms of process implementation (this is because when assessing the maturity level, the organization as a whole needs to be assessed and if it has wide variations in project implementations, then it requires more project assessments).

CMMI® assessments use a procedure described in SCAMPI based upon findings, observations and evidence to determine a maturity level. This is more complicated (requires more judgment) than the Capability Level rating mechanism in ISO/IEC 15504.

Difficulty consolidating data

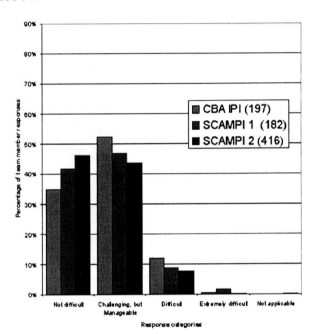

Difficulty consolidating CMMI® appraisal data.

Source: Software Engineering Institute. Appraiser Program Quality Report [70].

A Class A (e.g. SCAMPI^SM) assessment requires extensive and time consuming participation of a wide section of an organization. Generally certified Lead Assessors do not provide a maturity level report unless a Class A assessment is performed. Type B and Type C assessments require less time and effort but do not provide a maturity level rating.

Difficulty
maintaining a
realistic schedule
for the onsite
period

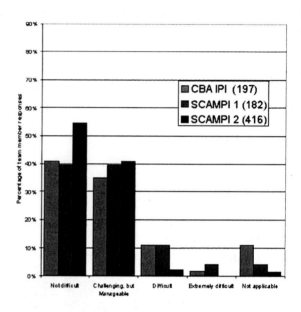

Maintaining a CMMI® schedule.

Source: Software Engineering Institute. Appraiser Program Quality Report.

The CMMI® Lead Assessor is totally responsible for the assessment result, even though the assessment team makes a collective judgment on each item in the assessment.

A formal CMMI® certification of an organization's maturity level does not exist. When an assessor submits an assessment result to the CMMI Steward, the steward only registers the assessment and does not certify that the Lead Assessor has correctly assessed the maturity level of the organization.

The SW CMM® introduced the Software Engineering Process Group (SEPG) and this has further been developed into Software Process Improvement Networks (SPINs) spanning multiple organizations using both SW CMM® and CMMI®. While Software Engineering Process Groups are not the only improvement team approach (and not always the best, especially when organizational improvement are needed), they have proven to be useful when supported by senior management in process oriented organizations. Readers are advised to read the chapter on process improvement for further information on this area.

CMMI® maintains a benchmarking database: Process Appraisal Information System, which has collected data on 1900 assessments since

1989. The analysis provided is at organizational level not process level, which is interesting for managers, but not useful for selecting process improvement projects.

ISO/IEC TR 15504 is about 340 pages long. With the latest revision of ISO/IEC 15504, the standard itself is much shorter and applicable to all process oriented standards, but readers may have to refer to Process Reference Model s from these other standards such as ISO/IEC 12207 or ISO/IEC 15288. The new part 5 is expected to be about 200 pages long.

ISO/IEC 15504-5 specifies work products in a generic manner so that they are consistently defined, but sometimes are too generic to be used without further knowledge or assistance.

ISO/IEC 15504 assessments are generally less time consuming than a Class A CMMI® assessment and assessors are better able to maintain the expected schedule. An ISO/IEC TR 15504-2 assessment is possible for the entire organization, for a single project or organizational unit or even a single process while still being able to provide a Capability Level result. Class A CMMI® assessments are always across the organization (cannot be tailored down).

The rating mechanism for determining Capability Levels uses a defined method partly based on simple mathematics, and partly based on the judgment of the assessors (each practice indicator may require assessor judgment but the summation of practice indicators is mathematical). This rating method is easier to use than that in SCAMPI.

ISO/IEC TR 15504-2 specifies a Competent Assessor for a conformant assessment, but in the case that a team makes collective judgment on the assessment result, the Competent Assessor can state this in the report. The assessment team or Competent Assessor can decide upon the how the assessment rating is made (individual, team, voting, consensus) involving the sponsor if desired.

The SPICE network operates the SPICE benchmarking server (hosted by ISS/PICS). Analysis is provided at process level with Capability Level ratings using the NPLF (Not, Partially, Largely, Fully) achieved rating scale. The benchmarks therefore are much more useful for selecting process improvement projects. The SPICE trial benchmarks are described briefly in Chapter 13.

ISO/IEC 15504-2 provides for multiple process reference and assessment models, while CMMI® specifies their own specific assessment model.

ISO/IEC 15504-2 is therefore more flexible and can be tailored to suit particular industry and business needs.

Author's conclusions.

SW CMM® [and] CMMI® have built up a solid following in some countries, particularly USA and India. When the business needs match these models, they are viable assessment models to adopt.

On the other hand, the flexibility of ISO/IEC 15504 both in terms of process coverage and assessment scope, plus its adaptability to specific industries is a strong advantage of the standard. The ability to perform assessments ranging in scope from one process in one project (or organizational unit) to a complete organization-wide assessment, means it is possible to perform very quick to very comprehensive assessments. In addition, as is described in the practical guide, the ability to use the capability scale independently of the process dimension to design higher capability business processes provides a powerful business advantage. The adoption of ISO/IEC 15504 is growing steadily as more industries realize these benefits.

For organizations looking for a general process assessment and improvement approach, either model is acceptable, but my preference would be to use ISO/IEC 15504 because of its assessment flexibility and the power of having various process reference models.

Annex 1 : ISO/IEC 15504 A Short History and Harmonization

The development of process assessment illustrates the improvement in understanding and application of process assessment for business purposes. Some key milestones in this development include:

➤ Research into process assessment techniques in IBM in the late 1970s early 1980s, including Ron Radice's Programming Process Study.

➤ The first publicly available process assessment method in 1987 [71].

➤ The start of the SPICE project in 1993 (which developed the early drafts of ISO/IEC 15504)

➤ The handover to an international standards working group of ISO/IEC DTR 15504 (Draft Technical Report) in 1997.

➤ The change from a software process assessment standard to a process assessment standard applicable to any process, in 2002.

➤ Ratification of the new international standard ISO/IEC 15504-2 in 2003.

The United States Department of Defence sponsored the Software Engineering Institute at Carnegie Mellon University to produce a process assessment standard for use in assessing Defence contractors. A preliminary version of the technical report was produced in 1987 [72]. This led to the production of the Software Capability Maturity Model. The Software Engineering Institute released the SW CMM V1.0 in 1991 [73].

From 1982, Bell were involved in process assessment methods (Bellcore Process maturity audit), and Bell Canada with Northern Telecom were investigating creation of process assessment standards to assess their major suppliers, eventually to release in 1991 the first version of Trillium [74]. In Europe there was a research group sponsored by the European Union that created the Bootstrap assessment methodology [75] in 1993-4.

In 1992, the United Kingdom Ministry of Defence through the Defence Research Agency sponsored a study called Improve-IT [76]. Based on study work, the British Standards Institution (BSI) proposed to that software process assessment be considered as an area for standardization and that the international community pool its resources to develop a standard for software process assessment, including the best features of existing software assessment methods [77].

The BSI suggested that a three part approach be used:

- A study period would be undertaken.
- Development of a draft international standard (Technical Report Type 2).
- Registration as a full international standard.

The ISO/IEC JTC1/SC7 WG10[42] study group met throughout 1992 and 1993 to specify the requirements for a common, international standard [78]. The new standard should be a software process assessment standard that provided both capability determination and a basis for process improvement, harmonize existing efforts, be flexible, consistent and reliable, yet be simple to use and understand. These requirements became the basis for starting the SPICE[43] project, which should then rapidly produce the software process assessment standard.

The SPICE project was administered by Alec Dorling with the project editor being Terry Rout of the Software Quality Institute, Brisbane, Australia. It began in 1993 and up until 1995 produced a series of draft documents to address the following requirements:

- Process assessments should be a basis for both process improvement and capability determination.
- It should be flexible with regard to applications, sectors, size, projects and organizations.
- Its coverage should encompass process, people and technology (products).
- Its outputs should be shown as profiles (process and capability).
- It should support existing standards such as ISO 9000 and ISO/IEC 12207.
- It should provide reliable and consistent assessment results.
- It should be simple to use and understand.
- It should be objective and provide quantitative results wherever possible.
- It should not be presumptive of specific organization structures, management philosophies, lifecycle models, technologies or development methods.

---

[42] ISO/IEC – International Standards Organization / International Electrotechnical Committee

[43] SPICE originally stood for Software Process Improvement and Capability Evaluation but due to French translation issues was changed to become Software Process Improvement and Capability dEtermination.

These requirements formed a basis for the conformance requirements of the standard (chapter 3) specifically in relation to the Process Reference Models, Process Assessment Models and the Capability dimension.

In addition to preparing documentation, the project also sponsored trials of the developing SPICE Process Assessment Model. The first phase trial occurred in 1995.

In 1995, the study group decided that an evolutionary approach would be taken to harmonize the software process assessment standard and ISO/IEC 12207 [79]. In addition, a reference model would replace the definitive process model, process descriptions would be restricted to more general statements of purpose, the capability dimension would become 'Process Capability Attributes' and there would be a section on conformance of process models. This was documented in the Kwa Maritime Agreement [80].

After the documentation was revised, the second phase trials occurred in 1996-1997, in which the author participated. At the conclusion of the second phase trials and inclusion of changes from the trials participants, the preliminary draft technical report was balloted. This consisted of nine parts [81] and was for a Software Process Assessment standard.

From 1997 to 2003, primary responsibility for the work was passed back to the joint ISO/IEC JTC1/SC7 WG10 working group. This group is responsible for the standardization process. This included a focus on the relationship to other standards and also transforming the draft and preliminary technical reports into what is now known as ISO/IEC 15504.

It is expected that the five part document set of the standard will be completed in 2005 (part 5 being the last part to be completed). In the meantime, users of the standard can use the earlier Technical Report version of the standard.

General Process Assessment standard

Because of harmonization with other standards (specifically ISO/IEC 12207, ISO/IEC 15288 and ISO 9000), ISO/IEC 15504 has become a general Process Assessment standard, and not just a Software Process Assessment standard.

It no longer specifies a single Process Reference Model, but specifies requirements for Process Reference Models. This allows the organization to select the most suitable existing model or create a model to suit their particular needs.

ISO/IEC 15504's development has been a worldwide collaborative effort, with input from twenty countries and many organizations from each country that are involved in the software and Information Technology industries, standards setting bodies, consultants and firms using and trying it within their own business.

Standards Development in ISO

The International Organization for Standardization (ISO) is a worldwide federation of national standards bodies from some 100 countries, one from each participating country. ISO's work results in international agreements, which are published as International Standards. ISO consists of many study and working groups.

Initially it was decided to create a Software Process Assessment standard, and it would be carried out under the auspices of a joint technical committee of the International Standardization Organization and the International Electrotechnical Commission. ISO/IEC JTC1 is the joint ISO and IEC technical committee, which deals with information technology. In 1993, this joint technical committee approved a new work item proposal, thus establishing working group WG10.

An early decision was made to pursue full standardization by first producing an International Organization for Standardization/International Electrotechnical Commission (ISO/IEC) Type 2 Technical Report (TR), and then converting it to a full international standard. As the subject of process assessment was still under technical development and there was the possibility of major changes subject to further agreement, the publication of a Type 2 Technical Report was deemed (by JTC1) to be more appropriate (than an International Standard).

The successive stages of the technical work for a Type 2 Technical Report were:

Stage 1 (proposal stage): A New Work Item Proposal was under consideration (1993).

Stage 2 (preparatory stage): A Working Draft was under consideration (1993-1995).

Stage 3 (committee stage): A Proposed Draft Technical Report (PDTR) was under consideration (1995-1996).

Stage 4 (approval stage): A Draft Technical Report (DTR) was under consideration (1996-1997).

Stage 5 (publication stage): A Technical Report (TR) was prepared for publication (1998).

Note: the reader will see reference to ISO/IEC TR 15504 throughout this book as some parts of the standard are from a Technical Report that are still current (planned update and issue in 2004-2005).

The decision to publish a Technical Report was taken by JTC 1 ballot on the Draft Technical Report in 1998.

Once the ISO/IEC TR 15504 document set was published, SC7 made a recommendation to JTC 1, stating whether the Technical Report should be revised and published as an International Standard. This revision has in fact been a major change from a 9 part standard in 1998 Technical Report focused on Software Process Assessment to a 5 part International Standard focused on Process Assessment.

This major change was due to the harmonization work occurring between the ISO/IEC JTC1 working groups responsible for ISO/IEC 15504, ISO/IEC 12207, ISO/IEC 15288 and ISO 9000.

The result is that the ISO/IEC 15504 was balloted in 2003 as an International Standard for Process Assessment. Due to the changes made in the past two years, it is now possible to use the standard with a wider variety of Process Assessment Models, covering the entire organizational process spectrum.

Note: a normative part of a standard is the formal part of the standard that must be complied with (i.e. it is mandatory), while an informative part of a standard is for information purposes or guidance and compliance is optional or voluntary.

ISO/IEC 15504-2 is a normative part of the standard, while ISO/IEC TR 15504-5 is an informative part. When the reader looks at the standard, they should determine which version they have (for example, the Technical Report version).

The ISO/IEC TR 15504 (Technical Report) informative parts are highly useful and form part of many implementations of the standard. Reference is made to these parts as they provide very useful guidance. The main Technical Report parts of general interest are:

ISO/IEC TR 15504-5 – the Process Assessment Model example.

ISO/IEC TR 15504-7 – the process improvement part.

ISO/IEC TR 15504-8 – the capability determination part.

Annex 2 : ISO/IEC CD15504-5 Proposed Process Assessment Model

As described throughout the book, the current part 5 of the ISO/IEC 15504 standard is still a Technical Report, its publication dating back to 1998.

The working committee is working on a major revision of this part with an aim to publish it in 2004. This will provide an updated comprehensive description of all the indicators including the base practices, work products, and the practice indicators for the process attributes at the various Capability Levels. Since the exemplar is the basis for several *Process Assessment Models*, it is important that users and model developers become familiar with the proposed changes (and comment on them when the review period occurs).

The revision will align part 5 with part 2 of the standard. I give the reader a quick look here at what can the new part 5 can be expected to look like[44]. In the document: ISO/IEC CD 15504-5.4 Information Technology — Process Assessment — Part 5: An exemplar Process Assessment Model, the major revision of the Process Assessment Model is based upon revisions to ISO/IEC 12207.

This will match the ISO/IEC 15504-2 Capability Levels to:

* Generic Practice Indicators (formerly called Management Practices),
* Generic Resource Indicators (formerly resource and infrastructure characteristics),
* Generic Work Product Indicators, and
* Related Process Indicators.

---

[44] No guarantees it will stay that way!

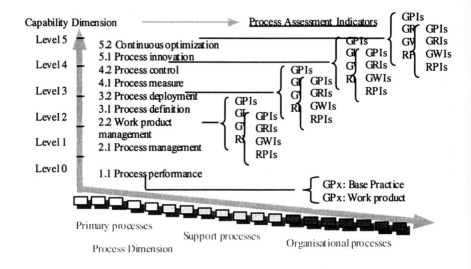

Capability Dimension Assessment Indicators.

I provide here a description of one Capability Level (level 3) as a guide to readers on what to expect the new descriptions to look like.

Level 3: Established process

The previously described Managed process now executes using a defined process that is based upon a standard process and that is capable of achieving its process outcomes.

The following attributes of the process demonstrate the achievement of this level:

PA 3.1 Process definition attribute

The process definition attribute is a measure of the extent to which a standard process is maintained to support the deployment of the defined process. As a result of full achievement of this attribute:

a) a standard process, including appropriate tailoring guidelines, is defined that describes the fundamental elements that must be incorporated into a defined process;

b) the sequence and interaction of the standard process with other processes is determined;

c) required competencies and roles, for performing a process are identified as part of the standard process;

d) required infrastructure and work environment for performing a process are identified as part of the standard process;

e) suitable methods for monitoring the effectiveness and suitability of the process are determined.

NOTE 1 A standard process may be used as-is when deploying a defined process, in which case tailoring guidelines would not be necessary.

Generic Practice Indicators for PA 3.1

GPI 3.1.1 Define the standard process that will support the deployment of the defined process.
• A standard process is developed that includes the fundamental process elements.
• The standard process identifies the deployment needs and deployment context.
• Guidance and/or procedures are provided to support implementation of the process as needed.
• Appropriate tailoring guideline(s) are available as needed.
GPI 3.1.2 Determine the sequence and interaction between processes so that they work as an integrated system of processes.
• The standard process's sequence and interaction with other processes is determined.
• Deployment of the standard process as a defined process maintains integrity of processes.
GPI 3.1.3 Identify the authorities, roles, responsibilities and competencies for performing the process.
• Process performance authorities, roles and responsibilities are identified.
• Process performance competencies are identified.
GPI 3.1.4 Identify the required infrastructure and work environment for performing the process.
• Process infrastructure components are identified (facilities, tools, networks, methods, etc).
• Work environment requirements are identified.
GPI 3.1.5 Determine suitable methods to monitor the effectiveness and suitability of the process.
• Criteria for process effectiveness and suitability are defined.
• Methods for monitoring the effectiveness and suitability of the process are

> determined.
> - Appropriate criteria and data needed to monitor the effectiveness and suitability of the process are defined.
> - The need to establish the characteristics of and/or trends in the process is considered.
> - The need to conduct internal audit and management review is established.

Generic Resource Indicators for PA 3.1

- Process modelling methods/tools;
- Training material and courses.

Generic Work Product Indicators for PA 3.1

11-00 Policy

- Provides evidence of organizational commitment to maintain a standard process to support the deployment of the defined process.

13-00 Process description

- Describes the standard process, including the fundamental process elements and appropriate tailoring guidelines.
- Addresses the performance, management and deployment of the process, as described by Capability Levels 1 and 2 and the process attribute PA 3.2.
- Identifies data and records to be collected when performing the defined process, in order to improve the standard process.

19-00 Repository

- Is used to support and maintain the standard process assets.

21-00 Standard

- Provides reference for the standards used by the standard process and identification about how they are used.
- Identifies and communicates the personnel roles and responsibilities for the project.
- Identifies the personnel performance criteria for the project.
- Identifies the tailoring guidelines for the project.

22-00 Strategy

- Identifies infrastructure and training needs to be satisfied.

Related Processes for PA 3.1

- MAN.2 Organization management;
- PIM.1 Process establishment;
- RIN.1 Human resource management;
- RIN.3 Knowledge management;
- RIN.4 Infrastructure;
- REU.1 Asset management;
- REU.3 Domain engineering;
- QUA.5 Audit.

The draft also contains a simple look-up table showing the related processes for each process attribute. It should be noted that some processes affect only one process attribute, while other processes are related to multiple process attributes, in some cases they affect multiple Capability Levels.

Annex 3: Glossary of Acronyms and Terms

assessed capability	the output of one or more recent, relevant process assessments conducted in accordance with the provisions of ISO/IEC TR 15504
assessment constraints	restrictions placed on the freedom of choice of the assessment team regarding the conduct of the assessment and the use of the assessment outputs
assessment indicator	an objective attribute or characteristic of a practice or work product that supports the judgment of the performance of, or capability of, an implemented process
assessment input	the collection of information required before a process assessment can commence
assessment instrument	a tool or set of tools that is used throughout an assessment to assist the assessor in evaluating the performance or capability of processes and in handling assessment data and recording the assessment results
assessment participant	Person of the organizational unit, who is interviewed by members of the assessment team for the acquisition of data on one or more processes to be assessed.
assessment purpose	a statement, provided as part of the assessment input, which defines the reason for performing the assessment
assessment record	an orderly, documented collection of that information which is pertinent to the assessment and adds to the understanding and verification of the *process profiles* generated by the assessment
assessment scope	a definition of the boundaries of the assessment, provided as part of the assessment input, encompassing the organizational limits of the assessment, the processes to be included, and the context within which the processes operate (see *process context*)
assessment sponsor	the individual, internal or external to the organization being assessed, who requires the assessment to be performed, and provides financial or other resources to carry it out
capability dimension	the set of process attributes comprising the capability aspects of the reference model of processes and process capability
conformant process assessment model	an operational model, used for performing assessments, which meets the defined requirements (for model purpose, scope, elements and indicators, mapping to the reference model, and translation of results) for conformance to the reference model
competent assessor	a person who has demonstrated the necessary skills, competencies and experience for performing process assessments
constructed capability	a capability constructed from elements of organizational units or of different organizations that are assembled for the purposes of achieving a particular specified requirement
defined process	the operational definition of a set of activities for achieving a specific purpose
enhanced capability	a capability greater than current assessed capability, justified by a credible process improvement programme
objective evidence	qualitative or quantitative information, records, or statements

	of fact pertaining to the characteristics of an item or service or to the existence and implementation of a process element, which is based on observation, measurement, or test and which can be verified
organizational unit	That part of an organization that is the subject of an assessment. An organizational unit (OU) deploys one or more processes that have a coherent process context and operates within a coherent set of business goals. NOTE: An organizational unit is typically part of a larger organization, although in a small organization, the organizational unit may be the whole organization. An organizational unit may be, for example: A specific project or set of (related) projects; a unit within an organization focused on a specific lifecycle phase (or phases) such as acquisition, development, maintenance or support; a part of an organization responsible for all aspects of a particular product or product set
practice	a software engineering or management activity that contributes to the creation of the output (work products) of a process or enhances the capability of a process
process	A set of interrelated activities, that transform inputs into outputs. NOTE: The term 'activities' covers use of resources (see ISO 8402:1994, 1.2). [ISO/IEC 12207]
process assessment	a disciplined evaluation of an organization's software processes against a model compatible with the reference model
process attribute	a measurable characteristic of process capability applicable to any process
process attribute rating	a judgment of the level of achievement of the defined capability of the process attribute for the assessed process
process capability determination sponsor	the organization, part of an organization or person initiating a process capability determination
process capability determination	a systematic assessment and analysis of selected software processes within an organization against a target capability, carried out with the aim of identifying the strengths, weaknesses and risks associated with deploying the processes to meet a particular specified requirement
process Capability Level rating	a representation of the achieved process Capability Level derived from the process attribute ratings for an assessed process
process Capability Level	a point on the six-point ordinal scale (of process capability) that represents the increasing capability of the performed process; each level builds on the capability of the level below
process capability	the ability of a process to achieve a required goal
process category	a set of processes addressing the same general area of activity
process context	the set of factors, documented in the assessment input, that influence the judgment, comprehension and comparability of process attribute ratings
process dimension	the set of processes comprising the functional aspects of the reference model of processes and process capability
process improvement	action taken to change an organization's processes so that they

	meet the organization's business needs and achieve its business goals more effectively
process improvement programme	al the strategies, policies, goals, responsibilities and activities concerned with the achievement of specified improvement goals. A process improvement programme can span more than one complete cycle of process improvement
process improvement project	any subset of the process improvement programme that forms a coherent set of actions to achieve a specific improvement
process outcome	an observable result of the successful implementation of a process
process performance	the extent to which the execution of a process achieves its purpose
process profile	the set of process attribute ratings for an assessed process
process purpose	the high level measurable objectives of performing the process and the likely outcomes of effective implementation of the process
proposed capability	the process capability that the organization proposes to bring to bear in meeting the specified requirement
provisional assessor	a person who has the skills and competencies to carry out assessments under the guidance and supervision of a competent assessor
software process assessment	A disciplined evaluation of an organization's software processes against a reference model.
software process	the process or set of processes used by an organization or project to plan, manage, execute, monitor, control and improve its software related activities
standard process	the operational definition of the basic process that guides the establishment of a common process in an organization
supplier	organization that provides a product or service to the customer
target capability	the process capability which the process capability determination sponsor judges will represent an acceptable process risk to the successful implementation of the specified requirement
work product	an artefact associated with the execution of a process
Architectural design	describes the division of the software into software components, their interfaces, the control structures and the implied mechanisms necessary to satisfy requirements. The architectural design can be documented in different views.
ECU	Electronic Control Unit - A computing device consisting of hardware, software, communication interfaces and locally connected sensors and actuators
Resource consumption of the software	specifies among others, the consumption of memory space and CPU time in sufficient detail
Release strategy	defines the sequence for realizing the software requirements according to the internal and external needs
Software	common name used for different types of software components such as application software or system software usually executed in an ECU. Software consists of software components. Software components consist of software units
Software component	An identifiable Part within the software that realizes a function

	on a particular ECU. The software component realizes vehicle functions (e.g. close sliding roof with transmitter key) as well as functions required ECU-internally (e.g. fail-safe-software)
Software design	super ordinate concept of architectural design, dynamic behaviour of the software and detailed design
Software requirements specification	Specifies the requirements for an instantiation of a function on a particular ECU
Software requirement	a single entity within the software requirements specification
Software units	The building blocks of software components. A Software unit is a small entity that can perform a coherent set of operations
Software integration test case	a test case used within the software integration to verify that the previously performed integration step has produced completely or partially integrated software that complies with the corresponding requirements
System	A system consists of several ECU's interconnected by a communication network for the purpose of realizing the coherent functions of a functional area
System design	the system design specifies the required sensors, actors, computing devices, internal and external interfaces, communication media and communication paths and the possible structuring into subsystems
System integration test case	a test case used within the system integration to verify that the previously performed integration step has produced a completely or partially integrated system that complies with the corresponding requirements
System requirements	The system requirements are specified by the customer and are among other relevant specifications: functional area requirements specifications, customer ECU requirements specifications, network specification
Test case	a description of the ordered steps necessary to decide upon the fulfilment of a requirement, together with the required preconditions and test setup (test object and test equipment). In case a corresponding verification criteria is defined, the test case contains a reference to it
Verification criteria	specifies the conditions under which the fulfilment and the non-fulfilment of a requirement can be objectively decided upon

Bibliography

[1] ISO/IEC 15504 standard consists of five parts:

ISO/IEC 15504-1 Information Technology — Process Assessment - Part 1 - Concepts and vocabulary

ISO/IEC 15504-2 Information Technology — Process Assessment - Part 2 - Performing an assessment

ISO/IEC 15504-3 Information Technology — Process Assessment - Part 3 - Guidance on performing an assessment

ISO/IEC 15504-4 Information Technology — Process Assessment - Part 4 - Guidance on use for process improvement and process capability determination

ISO/IEC 15504-5 Information Technology — Process Assessment - Part 5 - An exemplar process assessment model

[2] International Standard ISO/IEC 12207, Information Technology - *Software Life Cycle Processes*, International Organization for Standardization, International Electrotechnical Commission, 1995.

[3] International Standard ISO 9000:2000: Quality Management Systems: Fundamentals and Vocabulary.

ISO 9001:2000: Quality Management Systems - Requirements.

ISO 9004:2000: Quality Management Systems - Guidelines for performance improvements.

[4] ISO 9001:1994: Quality systems - Model for quality assurance in design, development, production, installation and servicing.

[5] Annex A: ISO 9004:2000: Quality Management Systems - Guidelines for performance improvements.

[6] ISO/IEC 14598:1998 Software Engineering – Product Evaluation. International Organization for Standardization, International Electrotechnical Commission.

[7] ISO/IEC 15939 Software Engineering – Software Process Measurement. International Organization for Standardization, International Electrotechnical Commission.

[8] ISO/IEC 12207:2002/Amd 2:2004, Information Technology - *Software Life Cycle Processes*, Amendment 2 International Organization for Standardization, International Electrotechnical Commission.

[9] Process purpose and outcomes definitions are taken from ISO/IEC 12207:1995/Amd 1:2002, Information Technology - *Software Life Cycle Processes*,

International Organization for Standardization, International Electrotechnical Commission, 1995. Amendment 1

[10] IEEE Std. 1517 – 1999 Standard for Information Technology – Software Life Cycle Processes – Reuse Processes.

[11] Carma McClure. Software Reuse – A Standards-Based Guide. IEEE Software Engineering Standards Series. ISBN 0-7695-0847-X

[12] ISO/IEC 15288 FDIS 15288:2002 (E) Systems Engineering – Systems Life Cycle Processes

[13] ISO/IEC 15288 FDIS 15288:2002 (E) Systems Engineering – Systems Life Cycle Processes

[14] V-Model Development Standard for IT-Systems of the Federal Republic of Germany, Federal Ministry of the Interior for the civilian Federal Administration. www.v-modell.iabg.de

[15] See ss01_01, www.itil.org, The Office of Government Commerce, British Government.

[16] The British Standards Institute: BS 15000:2000, Specification for IT Service Management

[17] John Torgersson, University of Boras. OOSPICE – The Road to Qualitative CBD. SPICE2002 Conference. www.oospice.com

[18] B. Henderson-Sellers, F. Stallinger, R. Lefever: Bridging the Gap from Process Modelling to process Assessment: the OOSPICE Process Specification for Component Based Software Engineering. Proceedings of the 28th EUROMICRO conference. Sept 2002, Dortmund, Germany. IEEE Computer Society. Loa Alamos, CA 2002.

[19] John Henderson-Sellers, COTAR UTS. The OOSPICE Project: Capability Assessment for CBD Methodology. SPICE 2001 Conference.

[20] Franck Barbier LIUPPA Université de Pau, France. Business Component-Based Software Engineering. Kluwer International Series in Engineering and Computer Science: Volume 705. ISBN 1-4020-7207-4

[21] Specifically the following standards were considered:

ISO 9000:2000 Quality Management Systems – Fundamentals and Vocabulary.

ISO 9001:2000 Quality Management Systems – Requirements.

ISO 9004:2000 Quality Management Systems – Guidelines to performance improvements.

[22] Application of Process Capability Determination to Quality Management. ESTEC Study Contract No. 14617/00/NL/CK. SYNSPACE AG, InterSPICE and Alenia Spazio, ESA.

[23] ISO/IEC TR 15504-5:1998 Software Process Assessment Part 5: An Assessment Model And Indicator Guidance

[24] ISO/IEC CD 15504-5.4 (2003) Information Technology — Process Assessment — Part 5: An exemplar Process Assessment Model

[25] Ann Cass, Christian Volcker, SYNSPACE AG, Paolo Panaroni, Intecs Systemi, Alec Dorling InterSPICE & Lothar Winzer ESA: SPiCE for SPACE: Creating an Assessment Method for Space Software Projects. SPICE 2000 Conference.

[26] ESA/ESTEC Study Contract No. 10662/93/NL/NB WO6-CCN5: ISO/IEC TR 15504 Conformant Method for the Assessment of Space Software Processes 14.7.1995

[27] Ann Cass, Christian Volcker, SYNSPACE AG, Paolo Panaroni, Intecs Systemi, Alec Dorling InterSPICE & Lothar Winzer ESA: SPiCE for SPACE: Creating an Assessment Method for Space Software Projects. SPICE 2000 Conference.

See also: Cass, A., Völcker C., et. al. SYNSPACE AG, "SPiCE for SPACE: A Process Assessment and Improvement Method for Space Software Development", *ESA Bulletin Number 107*, August 2001.

[28] Ann Cass, SYNSPACE AG, Juan M. Carranza ESA, Christian Völcker, Han van Loon, and Hans Stienen SYNSPACE AG: Identifying Risks with SPiCE: Experiences using R4S. SPICE 2002 Conference.

[29] ECSS standards included, among others: ECSS–E–40B Draft: Space Engineering – Software. ESA-ESTEC Requirements & Standards Division, Noordwijk, 28.6.2000.

ECSS-M-00A: Space Product Management - Policies and Principles. ESA-ESTEC Requirements & Standards Division, Noordwijk, 19.4.1996.

ECSS-M-00-03A: Space Product Management – Risk Management. ESA-ESTEC Requirements & Standards Division, Noordwijk, 25.04.2001.

ECSS-Q-80B Draft: Space Product Assurance - Software product Assurance. ESA-ESTEC Requirements & Standards Division, Noordwijk, 3.4.2000.

[30] Han van Loon, Robert Dietze, SYNSPACE AG, Fernando Aldea Montero, ESA, Software Reuse and SPiCE for SPACE. SPICE 2003 Conference.

[31] F. Stallinger, B. Henderson-Sellers, J. Torgersson: The OOSPICE Assessment Component: Customizing Software process Assessments to CBD. F Bartier (ed): Business Component-Based Software Engineering. Kluwer Academic Publishers. 2002.

[32] Gerhard Wagner, AUDI AG. SPICE im AUTOMOBIL. ASQF-02.11.2003 SPICE Days 2003 Nürnberg.

[33] Carnegie Mellon, Software Engineering Institute. Mary Beth Chrissis, Mike Konrad, Sandy Shrum. CMMI®: Guidelines for Process Integration and Product Improvement. Addison Wesley.

[34] http://www.sei.cmu.edu/cmmi/

[35] CMMI® for Systems Engineering, Software Engineering, and Integrated Product and Process Development (CMMI-SE/SW/IPPD, V1.1)

[36] Taken directly from section 7 of CMMI® for Systems Engineering, Software Engineering, and Integrated Product and Process Development (CMMI-SE/SW/IPPD, V1.1)

[37] EIA/IS 731 Electronic Industries Alliance. *Systems Engineering Capability Model (EIA/IS-731)*. Washington, D.C.: 1998.

[38] The Federal Aviation Administration Integrated Capability Maturity Model ® FAA-iCMM ® Version 2.0 An Integrated Capability Maturity Model for Enterprise-wide   Improvement.   Linda   Ibrahim   et   al.   September   2001. http://www1.faa.gov/aio/common/documents/iCMM/FAA-iCMMv2.htm

[39] The Federal Aviation Administration Integrated Capability Maturity Models (FAA-iCMM®) Appraisal Method (FAM). www.faaa.gov/aio

[40] Development Standard for IT-Systems of the Federal Republic of Germany. General Directive No. 250 "Lifecycle Process Model" (description of activities and products as an answer to the question of *"what"* will be done), General Directive No. 251 "Allocation of Methods" (description of the minimum requirements with regard to methods as an answer to the question of *"how"* it will be done), General Directive No. 252 "Functional Tool Requirements" (standardized criteria for the selection of tools which explain *"with what"* it will be done).

[41] The complete V-Model is available from IABG (Industrieanlagen-Betriebsgesellschaft.GmbH) in German at www.v-modell.iabg.de and in English at www.v-modell.iabg.de/vm97.htm#Engl

There is also an electronic process guide format at Fraunhofer IESE at www.iese.fhg.de/Vmodell/

[42] See section: The V-Model in the ISO and AQAP environment of General Directive No. 250 "Lifecycle Process Model".

[43] Davor Gornik, Rational® Software White Paper 2001. IBM ® Rational Unified Process: Best Practices for Software Development Teams

[44] Rational Software White Paper 2000: Assessing the Rational Unified Process against ISO/IEC 15504-5: Information Technology – Software Process Assessment Part 5: An Assessment Model and Indicator Guidance.

[45] A comparison of the IBM Rational Unified Process and eXtreme Programming. John Smith. 2003 Rational Software. IBM.

[46] Alistair Cockburn. Agile Software Development. 2002 Adison-Wesley, Pearson Education Inc. ISBN 0-201-69969-9.

[47] Kent Beck, First Class Software, Embracing Change with eXtreme Programming. IEEE Computer October 1999.

[48] Kent Beck, eXtreme Programming Explained: Embrace Change, Addison Wesley, Reading, Massachusetts, 1999. ISBN 0-201-61641-6

[49] Kent Beck, Test Driven Development – By Example. Addison Wesley, Reading, Massachusetts, 2003. ISBN 0-321-14653-0

[50] Kent Beck, Martin Fowler. Planning Extreme Programming. Addison Wesley , Reading, Massachusetts, 2000. ISBN 0-201-71091-9

[51] Hugh Robinson, Helen Sharp, Centre for Empirical Studies of Software Development, Open University, Milton Keynes, UK. XP Culture: Why the twelve practices are and are not the most significant thing. IEEE Proceedings of the Agile Development Conference 2003.

[52] Mark C. Paulk. eXtreme Programming from a CMM Perspective. IEEE Software vol.18 No. 6 November/December 2001.

[53] Michale K Spayd, QWEST Communications Inc, Evolving Agile in the Enterprise: Implementing XP on a grand scale. IEEE. Proceedings of the Agile Development Conference 2003.

[54] Donald J Reifer, XP and the CMM. IEEE Software May/June 2003

[55] Michele Marchesi, DIEE - University of Cagliari. Agile Methodologies and Quality Certification. 4th International Conference on extreme Programming and Agile processes in Software Engineering. www.xp2003.org.

[56] Control Chaos from ADM Inc, which is part of the Agile Alliance. www.controlchaos.com

[57] Christ Vriens, Philips Research – Software Engineering Services. Certifying for CMM Level 2 and ISO 9001 with XP@Scrum. IEEE Proceedings of the Agile Development Conference 2003.

[58] XP2003, 4th International Conference on extreme Programming and Agile processes in Software Engineering. www.xp2003.org.

[59] J. Nawrocki, B. Walter, A. Wojciechowski. Poznan University of Technology. Towards Maturity Model for extreme Programming.

[60] Software Quality Institute, Professor G. Dromey. Griffith University, Brisbane. Australia. www.sqi.gu.edu.au

[61] Director of Certification, International Assessors Certification Scheme (INT-ACS). Manchester, United Kingdom www.int-acs.org. See also the ASQF website for documents including www.asqf.de/asqf/documents/INT-ACS-100.pdf and www.asqf.de/asqf/documents/INT-ACS-4.pdf

[62] T. Rout, A. Tuffley, B. Cahill, Software Quality Institute. Griffith University CMMI Evaluation. Capability Maturity Model Integration Mapping to ISO/IEC TR 15504-2:1998. Contract No. 3726732, 28 March 2002.

[63] Christian Steinmann. HM&S GmbH. CMMI versus SPICE. Measurement Systems or Complementary Philosophies? June 2002.

[64] M. Paulk, B. Curtis, M.B. Chrissis et al. Capability Maturity Model for Software. Software Engineering Institute. CMU/SEI-91-TR-24, August 1991

[65]Software Engineering Institute, Capability Maturity Model® Integration, Version 1.1 for Systems Engineering, Software Engineering, and Integrated Product and Process Development (CMMI-SE/SW/IPPD, V1.1) Continuous Representation CMU/SEI-2002-TR-003 December 2001.

[66] Charles Weber, Description of CMMI Staged Model Representation. USAF Software Technology Support Center. 9 June 199.

[67] Software Engineering Institute technical report, Appraisal requirements for CMMI version 1.1. CMU/SEI-2001-TR-034 December 2001.

[68] Software Engineering Institute. Process Maturity Profile. CMMI® V1.1, SCAMPI^{SM} V1.1 Appraisal Results First Look. September 2003.

[69] Software Engineering Institute report on Process Maturity Profile for Software CMM CBA IPI and SPA Appraisal Results 2003 Mid-year Update. September 2003.

[70] Will Hayes. Software Engineering Institute. Appraiser Program Quality Report. November 2003.

[71] W.S. Humphrey and W.L. Sweet, A method for Assessing the Software Engineering Capability of Contractors. Technical Report CMU/SEI-87-TR23 September 1987.

[72] W.S. Humphrey and W.L. Sweet, A method for Assessing the Software Engineering Capability of Contractors. Technical Report CMU/SEI-87-TR23 September 1987.

[73] M. Paulk, B. Curtis, M.B. Chrissis et al. Capability Maturity Model for Software. Software engineering Institute. CMU/SEI-91-TR-24, August 1991.

[74] Coallier, F. (and others) (1991) The Trillium Model. Bell Canada.

[75] Koch, G. E. et al. 1993. Maturity Assessments: The BOOTSTRAP Approach.

[76] *ImproveIT*, 1991, ISO/IEC JTC1/SC7 N865.

[77] Proposal for a Study Period on Process Management, 1991, ISO/IEC JTC1/SC7 N872.

[78] Study Report: The Need and Requirements for a Software Process Assessment Standard, 1992, ISO/IEC JTC1/SC7 N944R.

[79] International Standard ISO/IEC 12207, Information Technology - *Software Life Cycle Processes*, International Organization for Standardization, International Electrotechnical Commission, 1995.

[80] WG10 Meeting Kwa Maritane: *Proposal for Changes to the Architecture*, 1995, ISO/IEC JTC1/SC7/WG10 N080

[81] ISO/IEC TR 15504 Technical Report consists of nine parts.

ISO/IEC TR 15504-1: Information Technology - Software Process Assessment Part 1: Concepts and Introductory Guide, 1996, ISO/IEC JTC1/SC7 N1592

ISO/IEC TR 15504-2: Information Technology - Software Process Assessment Part 2: A Reference Model For Processes And Process Capability, 1996, ISO/IEC JTC1/SC7 N1594

ISO/IEC TR 15504-3: Information Technology - Software Process Assessment Part 3: Performing An Assessment, 1996, ISO/IEC JTC1/SC7 N1596

ISO/IEC TR 15504-4: Information Technology - Software Process Assessment Part 4: Guide To Performing Assessments, 1996, ISO/IEC JTC1/SC7 N1598.

ISO/IEC TR 15504-5: Information Technology - Software Process Assessment Part 5: An Assessment Model And Indicator Guidance, 1996, ISO/IEC JTC1/SC7 N1601.

ISO/IEC TR 15504-6: Information Technology - Software Process Assessment Part 6: Guide To Qualification Of Assessors, 1996, ISO/IEC JTC1/SC7 N1603.

ISO/IEC TR 15504-7: Information Technology - Software Process Assessment Part 7: Guide For Use In Process Improvement, 1996, ISO/IEC JTC1/SC7 N1605.

ISO/IEC TR 15504-8: Information Technology - Software Process Assessment Part 8: Guide For Use In Determining Supplier Process Capability, 1996, ISO/IEC JTC1/SC7 N1607.

ISO/IEC TR 15504-9: Information Technology - Software Process Assessment Part 9: Vocabulary, 1996, ISO/IEC JTC1/SC7 N1609.

Index

A

Acquisition process, 67, 92, 93, 94, 103, 105, 110, 112, 135, 151, 153, 157, 177, 221, 225, 265, 266

agile development, 187

agile process, 188

Assessment Indicators, 129, 141, 143, 170, 171, 260

Assessment Planning, 35, 114, 149, 158, 178, 191, 192, 195, 197, 210, 220, 222, 224, 230

Assessment process, 18, 24, 26, 27, 33, 35, 39, 108, 126, 137, 183, 210, 230, 232, 233, 234, 235

Assessment purpose, 18, 33, 35, 36, 37, 38, 101, 109, 146, 170, 217, 265

Assessment rating, 31, 36, 44, 251

Assessment reporting, 38, 39, 55, 197, 210

Assessment Sponsor, 4, 26

Assessor judgment, 30, 37, 69, 70, 72, 131, 133, 170, 171, 204, 236, 249, 250, 251, 265, 266

Assessors, 4, 23, 26, 39, 40, 48, 54, 141, 170, 203, 205, 206, 208, 212, 217, 249

Attribute indicators, 131, 133. *See* Process Attributes

Automotive SPICE, 155

B

Business Process Mapping, 188

C

Capability Dimension, 18, 21, 43, 45, 128, 130, 142, 225

Capability Level, 18, 21, 43, 71, 73, 128, 131, 142, 184, 231

Level 0, 13, 21, 24, 46, 142, 198

Level 1, 13, 21, 28, 44, 47, 53, 71, 73, 75, 130, 132, 137, 142, 184, 185, 198, 220, 222, 230

Level 2, 13, 21, 44, 50, 53, 55, 71, 73, 75, 142, 156, 194, 197, 198, 199, 220, 222, 228, 230, 231, 236, 237

Level 3, 13, 54, 56, 57, 60, 64, 71, 73, 75, 142, 143, 144, 180, 184, 186, 220, 222, 227, 228, 229, 230, 232, 237, 260

Level 4, 13, 55, 57, 61, 64, 67, 71, 73, 76, 142, 186, 220, 222, 230, 231, 233, 234, 238

Level 5, 13, 21, 65, 66, 67, 68, 71, 73, 142, 199, 220, 222, 227, 230, 234, 239

Capability Level ratings, 71

Compatibility, 73, 210

Conformant process assessment, 1, 4, 6, 20, 24, 26, 28, 31, 33, 39, 79, 85, 124, 133, 198, 207, 227, 251

Conversion, 30

Crystal, 187

Customer, 4, 11, 53, 102, 103, 120, 121, 125, 135, 136, 153, 216, 225, 230, 231, 232, 233, 234, 235

About the Author

Han van Loon is a practicing management consultant, an Associate Professor at the International University in Geneva, Switzerland and a Visiting Professor at the Nottingham Trent University in Nottingham, UK.

He consults to companies wishing to improve their business results through process assessment and improvement.

Han has published articles on quality management topics over a period of more than 15 years, and worked with ISO/IEC 15504 since 1994. He has successfully led organizations to high quality process capabilities. Han adopted ISO/IEC 15504 and implemented improvements as an executive manager within CelsiusTech. He continues to advise and guide organizations in this field.

Han is an accomplished international speaker and has presented at conferences, led seminars and conducted training in Australia, Europe and USA. He maintains a web site at: www.starswebworx.ch/starsweb

The cover design is a combination of yin & yang, the symbol of infinity and the improvement cycle described in the book. It symbolizes the need for balance and tension, continuity and change, assessment and improvement, people and process.